T0190041

Lecture Notes in Intelligent Transportation and Infrastructure

Series editor

Janusz Kacprzyk, Systems Research Institute, Polish Academy of Sciences, Warszawa, Poland

The series *Lecture Notes in Intelligent Transportation and Infrastructure* (LNITI) publishes new developments and advances in the various areas of intelligent transportation and infrastructure. The intent is to cover the theory, applications, and perspectives on the state-of-the-art and future developments relevant to topics such as intelligent transportation systems, smart mobility, urban logistics, smart grids, critical infrastructure, smart architecture, smart citizens, intelligent governance, smart architecture and construction design, as well as green and sustainable urban structures. The series contains monographs, conference proceedings, edited volumes, lecture notes and textbooks. Of particular value to both the contributors and the readership are the short publication timeframe and the world-wide distribution, which enable wide and rapid dissemination of high-quality research output.

More information about this series at http://www.springer.com/series/15991

Octavian Iordache

Advanced Polytopic Projects

Springer

Octavian Iordache
Polystochastic
Montreal, QC, Canada

ISSN 2523-3440 ISSN 2523-3459 (electronic)
Lecture Notes in Intelligent Transportation and Infrastructure
ISBN 978-3-030-13166-1 ISBN 978-3-030-01243-4 (eBook)
https://doi.org/10.1007/978-3-030-01243-4

This Springer imprint is published by the registered company Springer Nature Switzerland AG
The registered company address is: Gewerbestrasse 11, 6330 Cham, Switzerland

Ascend with the greatest sagacity from the earth to heaven, and then again descend to the earth, and unite together the powers of things superior and things inferior....
Thus were all things created
Thence proceed wonderful adaptations which are produced in this way
Therefore am I called Hermes...

Emerald Tablet

Preface

High-dimensional structures understanding, designing, building, and operating are the objective of polytopic projects. Polytope is the general term of the sequence: point, segment, polygon, polyhedron …

Polytopic projects deal with emergent smart systems implementation. Emergence refers to the arising of novel and coherent processes or structures, during the self-organization in complex systems. The label smart, and interchangeably used-intelligent, refers to an object that was enhanced to new dimensions by implementation of communication and computational abilities. Devices in a smart factory should be able to operate autonomously according to the predefined or self-defined patterns or user requirements. Smart materials are materials that are capable of sensing the environment and actively responding to it in a controlled way.

The objective of polytopic projects is to confer on materials, objects, devices, technologies, or enterprises a new dimension, that of smartness, in addition to the dimensions they have.

Smart systems may be self-organizing, self-constructing, self-managing, self-repairing, self-aware, self-evolvable, and self-sustainable and may involve several other self-properties and self-processes at the level of being able to act on their own behalf, for a long period, like-alive, in a challenging environment of growing complexity.

Smart systems implementation should be envisaged giving that existing automatic tools, devices, methodologies, or organizations specific to third industrial revolution, reached or will reach soon their limits and new ways of design, construction, control, evolution, and problem solving are now manifestly required by the fourth industrial revolution.

Complex problems we face today challenge the Cartesian split situations as that between composition and decomposition, construction and deconstruction, synthetical and analytical methods, subject and object, and so on. The construction and deconstruction tendencies in knowledge and technologies, corresponding to synthesis and analysis, unifying and diversifying and to other antinomies should be reconciliated and correlated. That is because the interfaces where novelty flourishes and new information, new structures emerge, consist of correlated tendencies to unify and diversify. The antinomies specific to Cartesian split should stay together as complementary duality and synergy rather than conflicting, as inclusive rather than exclusive, as opportunities.

Since complementary descriptions may appear as basically incongruous, the conceptual and physical coordination of complementary descriptions needs a higher level description, a new dimension that emerges at a new hierarchical level of complexity. The more complex are systems, the higher dimensions, that is, polytopic architectures, are necessary for comprehension, designing, building, operation, and control.

This book proposes polytopic projects implementation as a post-Cartesian approach allowing facing high complexity. Polytopic projects are highly concentrated, quintessential ways of conveying knowledge, problems understanding and problem solving, developed, and implemented in concrete forms, as the amount of knowledge keeps growing and the subject areas we deal with are getting exceedingly complex.

Polytopic projects are grounded on a biologically inspired general framework shared by the functional organization of living organisms as informational and cognitive systems, the scientific and engineering methods, and the operational structure of existing smart systems.

In order to reflect different aspects, physical, technological, scientific, economical, and so on, the projects assign the polytopic character in the way we are looking for necessary messages into multipart objects that can be seen from many different perspectives, at multiple conditioning levels pursued in increasing and decreasing hierarchical order.

The book envisages describing the polytopic architectures that will help to combine several aspects of growing complexity systems into the emergent smart systems. The 4D approach developed in previous monographs dedicated to polytopic projects is continued here and extended to 5D and 6D architectures.

The book is divided into eight chapters. Chapter 1 outlines the limits of split in construction and deconstruction methods and justifies the need of specific coupling of both epistemological ways in technology and science. The 4D, 5D, and 6D polytope architectures are proposed as basic guides, condensed reference architectures, for understanding and solving problems, for designing, building, and controlling the emergent smart systems.

Chapter 2 is dedicated to integration and separation methods for technological schemes. Rooted trees and polytopic separations allowing multi-scale processing are presented.

Structuring and restructuring in supramolecular chemistry and quasi-species theory are presented in Chap. 3. This includes oriented synthesis and high-dimensional polytopes.

Chapter 4 outlines conditioning and randomness significance for processes. Forward and backward evolution is described and exemplified for mixing, growths, and restricted processes. Multi-scale transfer devices for processes intensification are proposed.

Chapter 5 examines assimilation and accommodation aspects for cognitive architectures. The role of development stages for cognition is outlined. Relation with the control and construction of hierarchies of skills is emphasized. Polytopic aspects of cognitive architectures are outlined.

Chapter 6 focuses on testing and designing correlation to data. Polytopic learning, pharmaceutical polytope, modeling-driven architectures, and elements of design hermeneutics are introduced.

Chapter 7 presents industrial systems based on additive and subtractive technologies.

The role of signed graphs is emphasized. High-dimensional printing and modular robots self-reconfigurations are presented.

Chapter 8 evaluates the perspectives and contains a retrospective of the domain for polytopic projects as physical and conceptual systems. Smart enterprises, smart factory, generic reference architectures, smart technologies, and smart society 4.0 and 5.0 projects are discussed.

Chemical engineering, material science, and systems chemistry are the preferred domains for examples highlighting the power of the polytopic projects for emergent smartness in high complexity conditions. Nevertheless, any sort of combinatorial problems that is characterized by a large variety of possibly complex constructions and deconstructions based on simple building bricks can be studied in similar ways.

The methodology proposed here focuses on the correlation between different domains and on polytopic projects as unified frameworks or general reference architectures for the analysis of reality and understanding in highly complexity systems and on practical implementations. The book will be useful to engineers, scientists, and entrepreneurs working in chemistry, material science, systems chemistry, pharmaceutics, biochemistry, environment protection, and ecology, to students in different domains of complex systems production and engineering, and to applied mathematicians.

Montreal, Canada
August 2018

Octavian Iordache

Contents

Abbreviations

CPS	Cyber Physical System
DGG	Dual Graded Graphs
DOE	Designs of Experiments
DPT	Dual Process Theory
FCA	Formal Concept Analysis
IIC	Industrial Internet Consortium
IIRA	Industrial Internet Reference Architecture
IVRA	Industrial Value Chain Reference Architecture
MDA	Model Driven Architecture
MVC	Middle Vessel Column
NBIC	Nano Bio Info Cogno
OMG	Object Management Group
PSA	Pressure Swing Adsorption
PSM	Polystochastic Model
PTP	Polytopic Project
RAMI	Reference Architecture Model Industry
RCA	Relational Concept Analysis
RDTE	Research Development Testing Evaluation
RSK	Robinson-Schensted-Knuth
SGAM	Smart Grid Architecture Model
SMB	Simulated Moving Bed

List of Figures

List of Tables

Chapter 1
Polytopic Projects

1.1 Polytopic Method

1.1.1 Bricks and Levels

To describe the world's reality Hartmann (1952) considered a hierarchy of four basic ontological levels "material or inanimate", "biological or animate", "cognitive or psychological", and "intelligent or spiritual" and emphasized the finite number of sub-levels to be taken into account at any basic level of reality.

Poli (2001, 2007) advocated the importance of levels or strata in the approach of formal ontology and distinguished three ontological strata of the real world: "material", "mental or psychological" and "social". These levels of reality or strata describe different classes of phenomena and are interdependent for example the social concept of trust depends on social entities which themselves interact in a material world. Levels of reality are characterized by the categories they use, and those categories imply a certain granularity, so that granularity appears as a derived concept. The ontological theory of levels considers a hierarchy of items structured on different levels of existence with the higher levels emerging from the lower but not reducible to the latter. The mental and the social strata are founded in the material stratum. This means that the categories and entities of the mental and social strata can be reduced to the category of material stratum, but only with a loss of information, so the reverse is not possible. Poli has stressed the need for understanding causal and spatiotemporal phenomena formulated within a descriptive categorical context for theoretical levels of reality.

Nicolescu's (2002, 2010) transdisciplinarity approach is based on three pillars: the levels of reality, the logic of included middle and the complexity. According to the logic of the included middle, in every relation involving two separate levels of experience, there is a third level that belongs simultaneously to both (Lupasco 1947). Complexity is the context in which this level of convergence finds place.

O. Iordache, *Advanced Polytopic Projects*, Lecture Notes in Intelligent Transportation and Infrastructure,
https://doi.org/10.1007/978-3-030-01243-4_1

A categorical mathematical model of the theory of levels intended for emergent biosystems has been elaborated by (Brown et al. 2007).

It was observed that complex systems exhibit hierarchical self-organization in levels under constraints. Self-organization will occur when individual independent parts in a complex system interact in a jointly cooperative manner that is also individually appropriate, such as to generate higher level organizations.

Complex systems can be observed and described at different levels of investigation.

Levels may be decomposed in sub-levels, in sub-sub-levels and so on (Kleineberg 2017).

In some cases this decomposition shows self-similarity and reversibility by composition. Subject for debate are the relations between n-levels and n-categories. The multi-scale, multi-discipline, multi-objective, multi-functions should be accounted for.

Consistent with Hartmann ontology, we refer to the four basic levels of reality: material, biological, cognitive and intelligent. The associated abbreviations are: Mat, Bio, Cogno, and Intel. Due to their specific role, the mathematical objects pertaining to the intelligent level have been highlighted under the term Math. Math is not a separate level but is included in Intel.

Polytopic projects start by asserting a goal, a problem to solve, a product to fabricate, an installation to design and control, a company to build and this is significant for the selection of building elementary bricks. For instance, if a chemical reaction is studied we can start the study at the molecular or atomic level but if a chemical factory is studied it could be better to start from devices or unit operations and evaluate their interactions.

Table 1.1 shows some examples of building bricks for different reality levels.

The building bricks are systems with different degrees of complexity (Velardo 2016).

Systems are content-independent and they transcend their particular domains. Systems are based on the relationships between different components that exchange information, rather than on the content of the information itself. A process can be described from a systemic point of view, even if the specific content of the process is unknown. As a consequence, systems can be used as high-level representations of specific processes.

Table 1.1 Building bricks

Levels	Building bricks
Mat	Electrons, atoms, radicals, processes, bricks, rocks, rivers, stars
Bio	Amino-acids, ribosome, cells, genes, organs, animals
Cogno	Letters, words, records, signals, sensations, concepts, designs, plans
Intel	Tools, devices, agents, programs, machines, companies, ideas, theories, societies
Math	Numbers, operations, graphs, propositions, equations, theorems, structures

For instance, it is possible to describe a molecule, a cell, a company or a society as a system. All items that make up reality are qualitatively equal, in the sense that they can all be regarded as systems, which exchange information with their environment. Reality, as a whole can be regarded as a system of systems, which exchange information with one another and which self-evolve over time.

Different systems in different domains manifest different degrees of complexity. For example, a cell is more complex than a molecule, a human is more complex than a cell, and an entire factory is more complex than a single tool. High complexity arises when moving from the physical domain, towards the biological, the cognitive and intelligent levels.

Table 1.2 shows some examples of levels in hierarchical order.

Coxeter (1973) defines polytope as the general term of the sequence: point, line segment, polygon, polyhedron, or more specifically as a finite region of n-dimensional space enclosed by a finite number of hyper-planes.

1.1.2 Duality and Complementarity

Dual or complementary pairs are those things, events and processes in nature that may appear to be contraries but are mutually related and inextricably connected (Kelso and Engstrom 2006). Both aspects of a complementary pair are required for an exhaustive account of phenomena. Dual aspects are dynamic, contextual, dialogical and relational.

Complementarity manifests itself in the whole-part behavior of reality, in the energy-informational properties of reality.

Some modern ideas about brain organization have emerged that may provide biological support for dual representation of the cognitive and intelligent processes.

Complementarity approach finds a strong foundation in the studies of metastable coordination dynamics of the brain (Kelso 2002; Kelso and Tognoli 2009).

Table 1.2 Levels hierarchy

Levels	Levels hierarchy
Mat	Atoms-molecules-supramolecules-supra supramolecules (vesicles) (Lehn 2007) States-processes-meta processes-meta meta processes (Iordache 2010)
Bio	Molec-organelles-cells-tissues-organs-organisms-populations-ecosystems
Cogno	Words-propositions-phrases-texts Objects/properties-concepts-FCA/RCA (Ganter and Wille 1999)
Intel	Devices-structures-suprastructures-supra suprastructures Real systems-sensors-controls-actions Components-communications-information-functions Texts-web 1.0-web 2.0-web-3.0 web 4.0 (Choudhury 2014) Data-models-meta models-meta meta models (OMG 2008) Manual-mechanical-electrical-electronic-digital industry (Schwab 2016)
Math	Points-lines-polygons-polyhedrons-polytopes (Coxeter 1973; Ziegler 1995)

Metastability has been highlighted as a new principle of behavioral and brain function and may point the way to a truly complementary neuroscience. Metastability is a result of a symmetry-breaking caused by the interplay of two forces: the tendency of the components to couple together and the tendency of the components to express their intrinsic independent behavior. The metastable regime reconciles the well-known tendencies of specialized brain regions to express their autonomy, that is differentiation or deconstruction, and the tendencies for those regions to work together as a synergy, that is integration or construction.

On a finer-grained scale, Grossberg (2000) has drawn attention to the dual, nature of brain processes and explained how are the brain functionally organized to achieve self-adaptive behavior in a changing world. Grossberg (2000) presents one alternative to the computer metaphor suggesting that brains are organized into independent modules, into parallel processing streams with dual properties. Hierarchical interactions within each stream and parallel interactions between streams create coherent behavioral representations that overcome the complementary deficiencies of each stream and support unitary cognitive experiences. Relation with parallel computing is obvious.

Cook (1986) model of the human brain provides another physiological basis for complementarity. Cook formulated the principle of dual control according to which systems pertaining to different levels of reality as the atom, the cell, the brain, the society and so on are isomorphic with respect to the duality of control mechanism. The duality neutrons-protons, DNA-RNA, right-left hemisphere, legislative-executive and many others have been outlined. Since dualities may appear to be intrinsically irreducible the conceptual integration or coordination of complementary descriptions requires a higher level description, a new dimension that emerges at a new hierarchical level of complexity. The emerging new dimensions are for the considered examples atomicity, life, mind, justice.

The more complex are systems the higher dimensional architectures that is polytopes, are necessary for comprehension, building and control. This isomorphism may be associated to the search for the general reference architecture in polytopic projects implementation.

A significant step in this direction is the complementarist epistemology and ontology as developed by (Ji 1995). Ji draws on the biology of the human brain, namely, the dual nature of its hemispheric specializations. The left and right hemispheres have relatively distinct cognitive functions, and the reality, as perceived and communicated by the human brain, is a complementary union of opposites (Ji 1995; Evans 2008).

Evans (2008) defines the System 1 and System 2.

Table 1.3 illustrates some of the hemispheric specialization of the human brain.

The dual aspects resulting from hemispheric specialization are significant. It suggests evaluating the correlation with mathematical tools as for instance, the dual graded graphs and Hopf algebras (Appendix A, Appendix B).

Generally speaking the Hopf algebra is both, the algebra and the dual of the algebra, the so-called coalgebra, which are compatible. There is a product rule which describes combining, composing, constructing, associating or synthesis of

Table 1.3 Hemispheric specialization

	Right Hemisphere	Left Hemisphere
1	Synthetic	Analytic
2	Intuitive	Rational
3	Implicit	Explicit
4	Fast	Slow
5	Concrete	Abstract
6	Analogical	Digital
7	Associative	Dissociative
8	Subjective	Objective
9	Spatial	Temporal
10	Parallel	Sequential
11	Holistic	Reductionist
12	System 1	System 2

the objects and a coproduct rule which describes breaking, decomposing, deconstructing, dissociating or analysis (Joni and Rota 1979; Dăscălescu et al. 2001).

The aspects recorded in first column in Table 1.3 are relatively close to the product rule (System 1) while the last column aspects are more close to the coproduct rule (System 2) (Evans 2008).

Duality in mathematics and physics is not considered as a theorem, but as a principle (Atiyah 2008). It has a simple origin and has a long history going back hundreds of years. It appears in many subjects in mathematics (geometry, algebra, analysis, combinatorics) and in physics (classical or quantum). Fundamentally, duality gives two different points of view of looking at the same object. There are many things that have two different points of view and in principle they are all dualities.

Duality studies reveal upward and downward, forward and backward, divergence and convergence, composing and decomposing, construction and deconstruction categorification and decategorification and other pairs of dual processes.

Table 1.4 shows some examples of duality pairs.

1.1.3 Closure

Closure concepts play a prominent role in systems theory where may be used to identify the whole system in correlation with its environment and to define the autonomy of the systems. Different closure concepts are linked to different facets of complexity.

A system is considered catalytically closed just in case every product of the system is also a catalyst in the system (Kauffman 1993, 2000). An autonomous agent must be an autocatalytic system able to reproduce and able to perform one or more thermodynamic cycles.

Constitutional dynamic chemistry is founded on a closure concept (Lehn 1999, 2002, 2002). Supramolecular chemistry is intrinsically a dynamic chemistry in view

Table 1.4 Duality

Levels	Duality
Mat	Wave-particle, fission-fusion, composition-decomposition, oxidation-reduction, separation-integration, stripping-rectifying, coagulation-dissolution, Contraction-expansion, thermodynamics-kinetics
Bio	Anabolic (associative)-catabolic (dissociative), phenotype-genotype, Metabolism-replication (Pross 2005), folding-unfolding, Material-symbol (Pattee 1995)
Cogno	Analysis-synthesis, coding-decoding, assimilation-accommodation (Piaget 1970), Concept-knowledge, general-particular, extent-intent, explicit-implicit
Intel	Assembly-disassembly, design-verification, convergent-divergent, Deterministic-stochastic, construction-deconstruction (Heidegger 1962), System 1-System 2 (Evans 2008), market-society
Math	Integration-differentiation, Hopf algebra (Sweedler 1969; Abe 1980), Dual graded graphs (Fomin 1994), categorification-decategorification

of the fragility of the interactions connecting the molecular components of a supramolecular entity and the resulting ability of supramolecular species to exchange their constituents. The same holds for molecular chemistry when the molecular entity contains covalent bonds that may form and break reversibility, so as to allow a continuous change in constitution by reorganization and exchange of building bricks. These features define a Constitutional Dynamic Chemistry (CDC) on both the molecular and supramolecular levels. On the molecular level, CDC is expressed in dynamic combinatorial chemistry (DCC) an approach that uses self-assembly processes to generate libraries of chemical compounds. In contrast to classical combinatorial chemistry which is based on vast collections of prefabricated molecules, DCC implements dynamic libraries via the continuous inter-conversion between the library constituents by recombination of their building bricks. Spontaneous assembly-disassembly of the building bricks through reversible chemical reactions virtually encompasses all possible combinations, and allows the establishment of adaptive processes owing to the dynamic interchange of the library constituents. The merging of the features: information and programmability, dynamics and reversibility, constitution and structural diversity, points towards the emerging self-adaptive, self-evolvable and smart chemistry. One may consider that CDC confers to chemistry an emerging new dimension that of constitution, in addition to the three dimensions of space (structure) and of time (kinetics). Thus, self-organization plays on a higher-dimensional chemistry to achieve complexification of matter and smartness (Lehn 2007).

Semantic or semiotic closure concepts are significant in theoretical biology.

According to Pattee (1995, 2001) biological organization consists of the correlation of two intertwined dimensions, which cannot be understood separately. On the one side, the organization realizes a dynamical and autopoietic network of mechanisms and processes, which defines itself as a topological unit, structurally coupled with the environment.

On the other side, it is shaped by the material unfolding of a set of symbolic instructions, stored and transmitted as genetic information.

The dynamical, that is, mechanistic and the informational that is symbolic aspects realize a distinct form of closure between them, which Pattee labels semantic or semiotic closure (Rocha 2001). This concept refers to the fact that while symbolic information, must be interpreted by the dynamics and mechanisms that it constrains, the mechanisms in charge of the interpretation and the material translation require that information for their own production. Semantic or semiotic closure, as a correlation or an interweaving between dynamics and information, constitutes an additional higher dimension of organizational closure of biological systems, complementary to the operational or efficient one. This higher dimension justifies the polytopic projects search for like-alive systems.

Autopoietic systems are also organizationally closed, that is they have a circular network of interactions, rather than a tree of hierarchically process (Maturana and Varela 1980).

The circularity and closure are associated also to the *Umwelt* concept that was introduced by von Uexküll in theoretical biology to describe how cognitive organisms perceive and interpret their environments. The *Umwelt* was defined as the part of the environment that an organism selects with its specific sense organs according to its needs (von Uexküll 1973). *Umwelt* theory asserts that a complex system doesn't responds to its environment but rather to its perception of the environment. A complex system actively creates its *Umwelt*, through repeated interactions with the environment. It simultaneously observes the world and changes it, the phenomenon which von Uexküll called a functional circle. The functional cycle is a cognitive architecture that self-organizes to accommodate the environment. The functional circle includes receptors and effectors. The sensory experience is based on interactions and these have specific purposes. The elementary unit of self-evolvable systems includes the functional circle of the following four parts: the environmental object, the receptors, the command generator and the effectors.

Circular reactions have been emphasized in the study of action scheme done by Piaget (Piaget 1970, 1971). Piaget called his general theoretical framework genetic epistemology because he was primarily interested in how knowledge develops in living organisms. Cognitive structures are patterns of physical or mental actions that underlie specific acts of intelligence and correspond to the stages of development.

The Piaget's action scheme, which constitutes the foundation of his learning theory, is a cycle including three elements: a recognized situation, an activity that has been associated with this situation, and an expected result. The recognition of a situation involves assimilation, that is to say, the situation must manifest certain characteristics which the organism has abstracted in the course of prior experience. The recognition then triggers the associated activity. If the expected result does not occur, the organism's equilibrium is disturbed and an accommodation may occur, which may eventually lead to the formation of a new action scheme. Accommodation does not take place unless something unexpected happens. Assimilation integrates new information in pre-existing structures while accommodation change and build new structure to understand new information. Piaget general equilibration theory offers a standpoint to consider the three level chains of interactions, namely biologic, cognitive and intelligent.

It can be observed that CDC, semantic closure, functional circle and circular reaction concepts have basic similarity despite the fact that they may refer to different levels of reality. They describe cycles or in other words loops of interaction between two or three successive levels or realms. Lehn focuses on molecular and supramolecular levels, Pattee focuses on two levels frameworks, the material versus biologic, or biologic versus cognitive, while von Uexküll and Piaget focuses on three level frameworks, material, biologic and cognitive level. It should be emphasized that some developments of the Piaget schemes embraces a four-level perspective to include intelligence level (Piaget and Garcia 1989).

(Bertschinger et al. 2006) elaborate upon the observation that cognitive systems can achieve informational closure by modeling their environment. Formally, a system is informationally closed if (almost) no information flows into it from the environment.

A system that is independent from its environment trivially achieves informational closure.

Philosophical hermeneutics answers that when we understand, it is because of the working of the hermeneutic circle (Heidegger 1962).

Hermeneutics can be considered a European correspondent to the American Pragmatist tradition in philosophy. Both offer a more relativist or constructivist understanding of knowledge, culture, practices, social interactions, in contrast to the dominant objectivist tradition in natural sciences. Hermeneutics has more explicitly dealt with issues of interpretation and meaning. The term hermeneutic goes back to Emerald Tablet of alchemy and to Hermes, the messenger between the "deities" and the humans. In order to be able to convey the will of the "deities" to the humans Hermes had to both be able to understand the original message and then able to translate it so that the intended meaning would be understood.

Table 1.5 illustrates the concept of closure.

1.1.4 Center and Rhythms

Centers are necessary to endorse the dialogue, coordination and coherence of the evolution on different levels.

The relation with concepts as "fold" (Deleuze 1988, 1993) "included middle", "hidden third" (Lupasco 1947; Nicolescu 2002, 2010) "tilde" (Kelso and Engstrom 2006) and "middle kingdom" (Latour 1993) is significant.

Following Lupasco, Nicolescu takes a transdisciplinary approach to nature and knowledge. Transdisciplinary knowledge, as based on the logic of the "included middle", necessarily leads to the inclusion of values and systemic understanding versus simple analytic reasoning. According to this logic, in every relation involving two separate levels of experience, there is a third level that belongs simultaneously to both.

Table 1.5 Closure

Levels	Closure
Mat	Balance, conservation, molecular networks, reactions network, Catalytic closure (Kauffman 1993, 2000) Constitutional dynamic chemistry, CDC (Lehn 2007) Algorithmic chemistry (Fontana 1991).
Bio	Homeostasis, cybernetics, hypercycle (Eigen and Schuster 1979) Semantic or semiotic closure (Pattee 1995, 2001; Rocha 2001) Closure to efficient cause (Rosen 1991; Mossio et al. 2009) Organizational closure (Maturana and Varela 1980) Semantic biology (Barbieri 2003)
Cogno	Umwelt (von Uexkull 1973), circular scheme (Piaget 1970, 1971) Informational closure (Bertschinger et al. 2006)
Intel	Hermeneutic cycle (Heidegger 1962; Ricoeur 1990, 1996; Gadamer 2004) Operational closure (Luhmann 1995), embedded systems, cyber-physical systems, industrial internet
Math	Axiomatic systems, circular-proofs (Blute 1996; Blute and Scott 1998)

If meta-theoretical background assumptions establish the framework for the disembodied mind, then other meta-theoretical assumptions must be found as a ground for framing an understanding of embodiment in which the lived body and the physical are complements, and not competing alternatives forever segregated into pure forms of reality. Such a set of assumptions is described by a perspective termed relationism (Latour 1993; Overton 2003, 2006, 2013).

Latour described this perspective as a move, a way from the extremes of Cartesian split to a center or "middle kingdom" position where all objects of knowledge are represented not as pure forms, but as forms that flow into each other across fluid boundaries. From this perspective, mind does not cause body, nor does body cause mind, nor do two entities interact; rather, mind and body are co-constituted, and as such, form non-dissociable complements. This is a post-Cartesian perspective (Overton 2013).

Table 1.6 shows some illustrations of the concept of center.

For polytopic projects the "Self" focus and coordinates the multi-level frames (Iordache 2017).

Modulated by the "Self", the swinging rhythms "up" and "down", forward and backward, separating and integrating, composing and decomposing, constructing and deconstructing, dialoging between the two complementary or dual ways are crucial for polytopic projects implementation.

Table 1.7 shows some examples of rhythms.

Since complementary descriptions may appear as antagonistic the conceptual and physical integration or coordination of complementary descriptions requires a higher level description, a new dimension that emerges as a new hierarchical level of complexity, as a center and a higher dimension in the same time. The more complex is a system the more dimensions and polytopic architectures are necessary to manage complexity.

Table 1.6 Center

Levels	Center
Mat	Nucleus, catalyst, template, tile
Bio	Enzymes, ribosome, ribotype, cellular nucleus
Cogno	Corpus callosum (Cook 1986), global workspace GW (Baars 2002) Global neuronal workspace GNW (Dehaene et al. 2003) Archetypal core, global landscape (Ehresmann 2012) Rich club (van Den Heuvel and Sporns 2011), dynamic core (Edelman 2003) Post-formal stages (Piaget 1971)
Intel	Abduction (Peirce 1956) Middle kingdom (Latour 1993) Relational metatheory (Overton 2003, 2006, 2013) Continuous engineering CE, cloud computing, market, business, web 4.0, "Self"
Math	Axioms, core arrays (Ceulemans et al. 2003), included middle, hidden third (Lupasco 1947; Nicolescu 2002, 2010)

Table 1.7 Rhythms

Levels	Rhythms
Mat	Periodic chemical reactions, linear-cyclic processes, flow reversal, cosmic
Bio	Hypercycle, biorhythms, meta-stability, chronotherapy, environmental rhythms
Cogno	Cerebral rhythms, problem solving, epistemological, circadian, development
Intel	Pressure Switch Adsorption, Simulated Moving Beds, economical, social
Math	Games, process semantics (Cockett 2006)

The swing between constructing and deconstructing steps implies supplementary dimensions. Correlating the constructing and deconstructing processes allows us make objects or run operations that could not be made or performed by only one of these processes in duality.

1.1.5 Generic Models

Processes, rather that just substances are targeted by the proposed modeling methods (Whitehead 1978). Consequently the time concept will play significantly.

Time and space are intuitive concepts and one cannot define their properties by entirely arbitrary mathematical rules. It is necessary to put in the frame, the physical and engineering knowledge allowing pragmatic and reasonable choice out of mathematical possibilities.

The adopted point of view in modeling is that the frames for time and space must agree first of all with the nature of phenomena and context of analysis of the studied system. Relational concepts of space and time should be implemented if the system analysis and synthesis can proceed on this basis.

On account of the change of space and time concept from universal to relational or contextual a generic model formulation, invariant under the choice of other than real field algebraic frameworks is envisaged.

Differential and integral models have been among the most fundamental tools for reductionist approach in science and engineering. To describe complex systems, instead of looking only at more and more detailed models and data, novel interpretations of these basic tools should be sought for higher levels of integration and differentiation.

The generic models, formally similar to the classical ones, capture the intuitions from the ordinary calculus and we have formally similar calculus rules of differentiation and integration. The generic models represent a new perspective of what it means to be differentiable or integrable in different contexts.

Modeling methods for systemic analysis in engineering have been developed in the frame of polystochastic models, PSM (Iordache 2010).

Polystochastic models, PSM, describe phenomena in which a family of conditioned processes, the so-called component processes produces a larger family of compound processes or in other words polystochastic processes, obtained by putting together the elements of the hierarchy of conditioned processes. As observed, polystochastic models, PSM, aim to focus on processes combination instead of substances combinations as in chemistry.

To account for complexity it was necessary to consider conditioned processes resulting as a process of processes or in other words a meta process. At the next level of conditioning, the meta meta processes have been considered and so on.

One of the modeling strategies was to maintain the pattern of the unconditioned level models but to adapt the definitions of parameters, functions and operations to the context.

In the reductionist Newtonian and Cartesian approach, time and space are considered independent of each other and of the existing entities and events occurring. Space and time are set beforehand and are invariant.

On the contrary, in Leibnizian or systemic approach, the space means order of coexistence, and time means order of succession. Leibnizian space and time are defining by referring to each other, as complementary and are dependent on the included entities and events, that is, on context (Deleuze 1988, 1993).

Replacing reductionist concepts for time and space, the relational concepts as, multi-dimensional, multi-scaled, dyadic, cyclic and multi-cyclic time and space are permitted and naturally implemented if the system analysis and synthesis can proceed on this basis.

The advantage is that we have structurally stable generic models admitting formulations invariant under the choice of the algebraic structure of numbers for space and time. The systemic models and methods recapitulate the real field models and methods for different contexts.

For example, the parameters and the functions of the real field conventional differential models for mixing have been replaced by other than real algebraic structures (Iordache 2011).

Statistical models should be adapted to the study of complexity in the frame of polystochastic models, PSM (Iordache 2012). Probability is the useful tool for representing uncertainty, conditioning, and information. It was observed that the set theory and corresponding probability theory are inappropriate frameworks to capture the full scope of the concept of uncertainty for complex systems. Uncertainty in set theory means non-specificity and exactly the specificity is important for complex systems. Conventional probabilities may be of interest when it is not detrimental to flat individual features but they are not adequate to account for strong individual deviations. Conventional probabilities are inappropriate to illustrate qualitative concepts as fuzziness, similarity, partial truth and possibility, all having significant role in complexity studies.

Answering to the need of probability-like concepts in the study of complex multi-scale systems, the probability construction for different algebraic frame has been considered.

For combinatorial structures, informational probabilities and informational distances based on similarities and possibilities have been defined. This allows significant simplification of calculus. Several informational entropy criteria allow characterizing self-evolvability trends at different complexity levels. Such criteria concern entropy production and self-evolvability (Iordache 2012).

1.1.6 General Theories

The connection of the approach proposed here with other general theories will be highlighted in this paragraph.

Table 1.8 shows some examples of developed general theories.

In the domain of sciences of mater there exists a highly developed theory, the quantum chemistry allowing the successful classification of the elements in the chemical periodic table. The theoretical explanation of the chemical classification, by Schrödinger's wave equation and Pauli's exclusion principle, is a success of physics and chemistry in the twentieth century. One can look at this theory from very diverse points of view that follow closely the above presented polytopic projects methodology. The building bricks are the electrons. The levels refer to discretization of the energy of oscillators and the discretization of angular momentum.

De Broglie and Schrödinger approaches, emphasized the wave-matter duality.

Finally, these diverse points of view are all corollaries of that of Heisenberg overpowering statement: physical quantities are governed by non-commutative algebras.

Chemists working at the material level of reality make use of theoretical tools as the periodic table for classification, balance equations describing transfer phenomena in engineering and models as the wave equation from quantum chemistry. Also the chemists have many devices and technologies to combine substances and afterwards to decompose and separate them.

Table 1.8 General theories

Levels	General theories
Mat	Quantum theory, Self-organization theory
Bio	Hypercycle (Eigen and Schuster 1979) Complementarity (Kelso and Egstrom 2006) Cybernetics, Biosemiotics (Barbieri 2003)
Cogno	Genetic Epistemology (Piaget 1971), Dual Process Theory, DPT (Evans 2008) Reference Architectures in Industry 4.0: RAMI 4.0, IIRA, SGAM, IVRA
Intel	General System Theory, Ontology (Hartmann 1952) Epistemology of Deconstruction-Reconstruction (Heidegger 1962) Philosophical Categories (Peirce 1956) Cyber-Semiotics (Brier 2008, 2009) Polytope Projects, PTP (Iordache 2012, 2013, 2017)
Math	Meta Meta Model (OMG 2008) Differential Linear Logic and Integral Linear Logic (Ehrhard and Regnier 2006; Blute et al. 2010) Hyperstructures (Baas 2009, 2015a, b), n-Categories, Memory Evolutive Systems, MES (Ehresmann and Vanbremeersch 2007)

The challenge for researchers focusing on other levels of reality, biological, cognitive, industrial, socio-economical, logical and mathematical systems is to develop analogous methods, techniques and models for coding and classifying, devices for composing and decomposing specific building bricks and components.

Self-organization is a core concept of systems science. It refers to the ability of a class of systems, the self-organizing systems, to change their internal structure and their function in response to external circumstances. Some elements of self-organizing systems are able to manipulate or organize other elements of the same system in a way that stabilizes either structure or function of the whole against external fluctuations. The process of self-organization is often achieved by growing the internal space-time complexity of a system and results in layered or hierarchical structures or behaviors. This process is understood not to be directed from outside the system and is therefore called self-organized.

Modern ideas about self-organization start with the foundation of cybernetics in the 1940s. Later, the concept was adopted in physics and nowadays pervades most of natural sciences. Some theories focused the processes of self-organization in systems far from equilibrium (Nicolis and Prigogine 1977; Haken 2000).

Chaos theory and fractal theory (Mandelbrot 1982; Coppens 2004) was a line of inquiry into nonlinear systems in mathematics and engineering. Autopoiesis and self-maintenance were at center stage in biology (Eigen and Schuster 1979). The theory of autocatalytic hypercycles supposes that organisms consist of functionally related self-replicative units formed into multiple feedback loops.

Self-organizing systems have assumed center stage in the cognitive science (Maturana and Varela 1980) and social sciences (Luhmann 1995).

Engineering is beginning to use of the concept of self-organization in connection with the approach of nanotechnology applications and the growing complexity of human artifacts (Brueckner et al. 2006). Self-reconfiguring automata, is a recent domain of research for engineers (Yim et al. 2007; Gilpin and Rus 2010, 2012).

In *On a New List of Categories*, Peirce formulates a theory of categories that can demonstrate what the universal conceptions of reality and of thought are (Pierce 1956).

Peirce's categories are meant to provide a basis for an exploration of a large variety of phenomena, including natural, biological, technological and reasoning.

Peirce proposed an initial list of five philosophical categories: substance, quality, relation, representation and being.

Later, Peirce discarded substance and being from his initial list of five categories and focused mainly on quality, relation and representation which he called in his technical terms firstness, secondness and thirdness, respectively. These have been associated to 1-level, 2-level and 3-level systems (Iordache 2012).

Taking inspiration from Peirce's philosophy, Brier formulated a transdisciplinary theory of information, semiotics, consciousness and cultural social communication illustrated by the four fold cybersemiotic star (Brier 2008).

The four folds of the star correspond to the four main areas of knowledge that is: Material, Living, Consciousness and Mentality. A comparison with the Hartmann's ontological hierarchy is of interest (Hartmann 1952).

The center of the cybersemiotic star was associated by Brier to semiotic mind. It may be considered as a meta-representation of the four fold star and it is linked to self-evolvability and smartness concepts.

The polytopic projects, PTP, have been proposed as general theories, basic cognitive architectures, for problems understanding and solving, for designing, building and managing emergent smart systems (Iordache 2012, 2013, 2017).

In all human activities we consider building bricks and collections of bricks.

We study how we can make new objects and collections out of old ones by using their properties, relations and interactions. In order to do so it is important to have in mind what kind of architectures to use when forming new structures and organizations out of given collections of objects. The main purpose is to point out that there is an accumulation of high-dimensional architectures waiting to be explored.

The polytopic projects, PTP, describe the general reference architecture allowing facing complexity as encountered in polystochastic models, PSM advances.

The polytopic projects start from a biologically inspired general architecture, useful for artifacts building, information representation, design, operation and calculus.

Highlighting different aspects, material, technological, scientific and socio-economical, the resulting architectures will be also interesting in themselves as geometrical objects like lattices, hypercubes and other polytopes.

The issues raised by implementing polytopic projects concern the hardware and software, the foundational brain-like machine structure, the engineering methods and so on.

The polytopic projects are based on recent findings from material science and electronics, biology, psychology and informatics and it is expected to provide a general framework for higher level innovative quantitative and theoretical research in these domains.

The polytopic projects follow a natural trend to unify and standardize the research discovery, design and control methods (Langley et al. 1987; Langley 2006).

Notable theoretical perspectives more or less close to the polytopic projects PTP, approach are the hyperstructures (Baas 2009, 2015a, b, Baas et al. 2004) and the memory evolutive systems MES (Ehresmann and Vanbremeersch 2007). These research directions develop, in different ways, the idea of iterative constructions of systems of systems over systems and so on. The systems of different levels of reality show specific properties.

Baas et al. (2004) compare two different approaches to the modeling of complex natural systems, in particular of their hierarchical organization with higher-order structures and their emergence processes. These approaches are the hyperstructures and the memory evolutive systems, MES. The hyperstructures are structural while memory evolutive systems based on category theory, take dynamics more into account. The dynamical organization and mechanisms developed for memory evolutive systems rely on general ideas that might be disengaged from the categorical setting and extended to the general frame of hyperstructures.

1.2 Polytopic Framework

1.2.1 Unifying and Diversifying

The domain of complexity concerns the problems that can be seen in nature, industry and society and use to be considered as very hard. This includes problems as for example drugs design or drugs delivery, environmental experiment organization and data analysis, cognitive architectures implementation, traffic control and security, market evolution, balance in biosphere, society organization and so on. We tend to throw up our hands at these problems, thinking that they are just too complicated and that individually, on the basis of one point of view and one scientific discipline, we can't make a difference.

What many of these problems have in common, actually, is that they exhibit a hierarchy of combinatorial emergent patterns and levels caused by the local and global interactions of a large number of individual parts, aspects, solutions and perspectives. Emergent patterns and conditioning levels compose and decompose according to some internal laws. This produces a combinatorial explosion of new patterns and levels due to large number of subsequent moves in the composition-decomposition game.

There is a strong request for scientific tools to think reliably and systematically about such hard problems.

Modern technology and research would be unthinkable without interdisciplinary and transdisciplinary approaches. Consider for example the development of chemistry as a source of chemical disciplines as physical chemistry, chemical technology, chemical engineering, biochemistry, material science, systems chemistry, chemostatistics, computational chemistry, and chemical informatics. All these cover aspects of chemistry that are influenced by at least one other scientific discipline (Bonchev and Rouvray 2005; Faulon and Binder 2010). The practical problem is how the converging and diverging technologies and methodologies fit and develop into the existing, interdisciplinary framework of modern chemistry research?

Technologies are based on science that has found a particular application area, where the driving force is a practical outcome. Different application areas can have the same scientific principles and approaches in common and only differ by the details and specificities of their implementation. The convergence and divergence of technologies provides complementary benefits to interdisciplinary research.

The way by which technology in one application area is transferred to another application area should be based on the development of simplified and unified frameworks. The use of structural analogies provides a link to scientists, engineers and entrepreneurs since structural analogies allow individuals to grasp concepts directly by relating them to something already known. Thus, such unified and essential framework may have significant technological and social impact because they may provide direct means of communicating transdisciplinary research using structural analogies.

Confronted with the differentiation of disciplinary knowledge, it is difficult for any specialist to understand more that a fraction of his specialized domain. The management of the cooperation of different disciplines and points of view for complex problem solving is the current concern. It is necessary to find new ways to radically simplify and unify knowledge, to propose new paradigms capable of dealing with a collective research by communities for concrete solutions to a problem having no solution known with certainty but multiple solutions and achievability ways (Kaku 1999).

Complexity is the research field emerging around the conviction that some problems pertaining to different domains as for instance, material science, molecular biochemistry, neuroscience, computer science, telecommunications, manufacturing and economy can be challenged scientifically in a unified way, by means of which progress in understanding aspects in either field can be fruitful to the others. By integrating disparate fields, we may link very different disciplines that can learn and benefit from one another.

The process of finding unifying principles either at the microscopic or macroscopic levels of complex systems engineering, is delayed by the problems of technical language where different concepts share overloaded names while similar concepts may have different names (Buchli and Santini 2005).

Despite substantial knowledge about complex systems the application of this knowledge to the practical domains remains difficult. Efforts to manage complexity are still scattered over many scientific and engineering disciplines. Attempts to establish complexity engineering as a discipline are hindered by misunderstandings over basic terms such as for instance emergence and causation. Although terminology standardization is a necessary feature of communication it can also pose a barrier impeding the technological progress.

As the amount of knowledge keeps growing exponentially and the subject areas we deal with are getting exceedingly complicated, highly concentrated, quintessential ways of conveying knowledge should be developed and implemented.

It was argued that a potential incompatibility between techno-scientific and socio-economical approaches to ever growing complexity problem may develop.

Therefore, a broader framework is needed to encompass multiple aspects since we need to make coherent theories and models for different subject areas.

The proposed methodology focuses on the relation between several domains and on polytopic projects, PTP, as a general unified framework, a cognitive and modeling architecture, for the analysis of truth and meaning in highly complex problem understanding and solving.

For reductionism, the reference is to the classical Cartesian and Newtonian assumptions that the dynamics of any complex system can be understood from studying the properties of its parts. Complex systems are therefore broken down into their components and each piece is studied individually by way of disciplinary and sub-disciplinary approaches. The challenge is to find the entry points from where to address the particulars of the system.

It is expected that once one knows the parts, the dynamics of the whole can be derived.

If we consider the case of chemical engineering, as for the models themselves, in the reductionist or analytical approach to modeling, engineers try to represent a whole chemical plant by decomposition into smaller units focusing down to devices, particles, crystals, droplets, bubbles, and finally to molecular processes, going into more and more detail. The basis for this reductionist approach lies in the first principles of mass, energy, momentum, and population balances. Such modeling attempts are necessary and should be continued, but not enough attention has been paid to systemic models based on an integrated approach that considers the global behavior of complex systems as a whole.

The reductionist approach is a useful method for gaining knowledge about certain aspects of the natural or artificial world and for supporting the development of technology precisely because it selects phenomena that have a simple explanation and control. But the cost of restricting to simplicity only is to be unable to represent the full range of possibilities offered by systems engineering. As a mandatory way of success of thinking and a firm base for undertaking, the reductionism should be complemented by the way of systemic or integrative thinking.

The basic assumption underpinning the integrative approach is that the properties of the parts contribute to our understanding of the whole, but the properties can only be fully understood through the dynamics of the whole. The research focus in integration is on the relationships between the components, that is, on interconnectedness, interdependencies and interactions. In this viewpoint, the whole is different from the sum of its parts. Consequently, breaking complex systems down into their individual components by the methods of reductionism is just a one way approach of the truth, and while it may afford many useful insights, it is necessary to put the pieces together again by way of integration. Instead of looking at more and more details and data, novel first principles should be sought at higher levels of integration. The call is for multi-disciplinarity and for bringing the multiple specialties contained in disciplines together in what can be labeled disciplinarily.

Despite the differences in focus there are no automatic or necessary contradictions between the two basic ways of knowledge. The one focuses on the properties of parts, the other on the relationship between them. Put together, they stand out as complementary duality and synergy rather than conflicting, as inclusive rather than exclusive, as opportunities.

A continuous process of unification, differentiation and re-unification is the methodology endorsed here. The mix and remix of both ways, the reductionist and the systemic ways, that is, the differentiation and unification ways, is the suitable method to confront high complexity advent.

While, in some cases, traditional engineering studied the ways of taking the complexity out of the systems, we now have to allow complexity to come back in, to complement the reductionism and learn to exploit as oportunities the epistemological ways of unification and diversification for our own good (Kelso and Engstrom 2006; Longo et al. 2012).

Unaccompanied, the divergence or deconstruction way as well as the convergence or construction way does not quite grasps the essence of creativity required by emergent smartness. Tendencies to diverge should interplay the tendencies to converge and it is the mix and remix of both that matters for high complexity management (Bainbridge and Roco 2006; Sharp and Langer 2011).

1.2.2 Advanced Polytopic Framework

The polytopic projects are based on a wide-ranging biologically inspired architecture, useful for designs, operations and calculus, artifacts building, knowledge representation and development (Iordache 2012, 2013, 2017).

The projects assign the polytopic character in the way we are looking for necessary messages into essential objects that can be seen from many different perspectives. Trying to reflect different aspects, physical, technological, scientific and socio-economical, the resulting architectures would be also interesting in themselves as geometrical objects like n-cubes, posets, lattices and polytopes.

The issues raised by polytopic projects concern the groundwork of material or device structures, the hardware and software, and also the considered necessary scientific and engineering methods.

The evolution of reference modeling architectures for growing complexity study is shown in Fig. 1.1. An evolution from 0D to 6D polytope is considered here.

It highlights the trend towards n-cubes representations.

Figure 1.1a corresponds to the first level of modeling, for instance, that of examination and description of the states and conditions as separate entities. It shows sets of dots or 0D cubes. The next step is that of relations or mappings between states and conditions as encountered for instance in learning models (Fig. 1.1b). This is shown as a line or 1D cube.

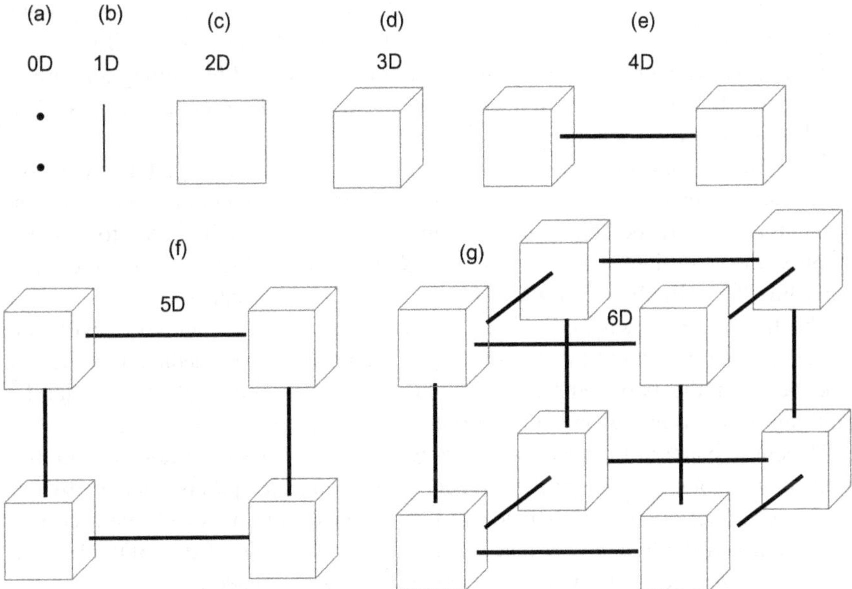

Fig. 1.1 Reference modeling architectures for growing complexity

The next steps in modeling engage several levels of conditioning and evolvability. The evolvability requires more than learning. It needs innovation that is the capacity for change to autonomous walk into new life cycles, for instance new type of products, new market niches, new organizations and new levels. This entails a high-dimensional perspective and also the systems closure as shown in Fig. 1.1c. This is shown as a square or in other terms a 2D cube. Observe that to pass from 1D to 2D one needs doubling followed by relating of the resulting two segments.

Four levels systems corresponding to states, conditions, meta conditions and meta meta conditions have been highlighted in different engineering domains were evolvability features were studied.

Four levels systems corresponding to states, processes, meta-processes and meta meta processes have been studied under the name of polystochastic models PSM (Iordache 2010, 2017).

Figure 1.1d shows a 3D cube. The front face, a 2D cube, is complemented by the back face, another 2D cube, and the correspondent corners are related. Once again we are confronted with doubling followed by relating. Figure 1.1d illustrates both the evolution as the involution associated to the front and back faces of the 3D cube.

The continuous grows of complexity imposed a new paradigm that of transition from evolvability to self-evolvability, that is to systems that self-configure, self-optimize, self-control, self-heal, self-direct and so on, based on a set of higher-level intrinsic capabilities and meeting of the user-specified variable objectives.

The polytope shown in Fig. 1.1e illustrates the architecture of potentially self-evolvable systems. It is a representation known as hypercube or 4D cube and consisting from a pair of 3D cubes. The 4D cube is obtained by joining all corners of the first 3D cube with the corresponding corners of the second 3D cube. It is doubling followed by relating.

The 8 lines connecting the two 3D cubes are represented in Fig. 1.1e by a thick hyper-line. In this way the 4D cube has a simplified representation. All dimensions higher than 3 may be represented using thick hyper-lines in order to visually simplify the resulting structures (Thiel 1994; Ammon 1998). Using the thick hyper-line the 4D cube is shown in Fig. 1.1e as a "line of cubes".

The first cube shown in Fig. 1.1e would coordinate the four level evolvable frames shown on different faces of the second cube. Swinging, mediated by the first cube between the two complementary faces of the second cube is crucial for reuniting different trends and self-evolvability.

The entire pattern of representation shown in Fig. 1.1 is significant for cognitive architecture evolution. A segment is made by two connected points, a square by two segments connected, a cube is made of two squares, front and back, with corners connected, and the hypercube, that is the 4D cube, is made of two 3D cubes, the first and the second, again with corresponding corners connected.

This doubling pattern continues for higher dimensions, but it is more difficult to draw and to interpret.

Since there is no fixed limit for growing complexity, high-dimensional polytopes as for instance 5D and 6D cubes and others polytopes may be considered as reference architecture instead of the 4D cube. Figure 1.1f shows a 5D cube. Figure 1.1g shows a 6D cube. The internal cubes again should have the corresponding corners connected as for the 4D cube. Using the thick hyper-lines, the 4D cube is shown as "line of cubes". Using the thick hyper-lines, the 5D cube is shown as "square of cubes" while the 6D cube is shown as "cube of cubes" (Seitz 1985).

Figure 1.2 shows different but equivalent presentations of the 4D cube from Fig. 1.1 e.

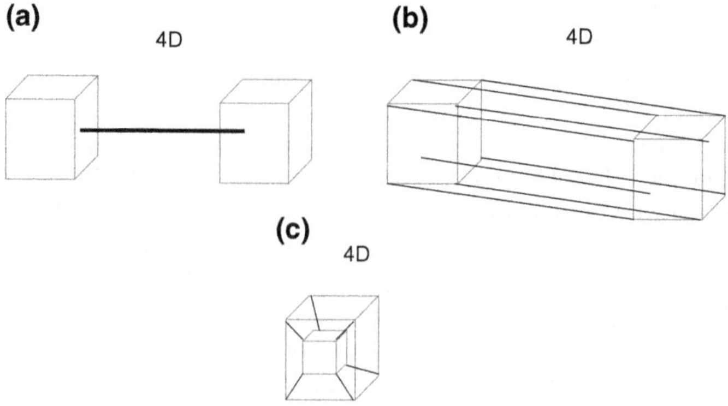

Fig. 1.2 Different presentations of the 4D cube

Figure 1.2a shows two 3D cubes connected by a thick hyper-line. This thick hyper-line is a simplified representation of 8 connections of the corresponding corners as shown in Fig. 1.2b. In Fig. 1.2c the first cube is introduced in the second cube. The 8 connection between corresponding corners are still present. The three presentation of the 4D polytope shown in Fig. 1.2 may be considered as equivalent.

Figure 1.3 outlines more details for the polytope shown in Fig. 1.2c (Iordache 2013, 2017).

Usually S, K1, K2 and K3 are associated to the 0, 1, 2 and 3-levels while S', K1', K2' and K3' are associated to the modified 0, 1, 2 and 3-levels.

The polytope outlines that after the converging, constructing or direct way S → K1 → K2 → K3 we need to look at the diverging, deconstructing or indirect way K3′ → K2′ → K1′ → S′. S and S′ were denoted also by K0 and K0′ in some case studies.

The swinging from direct way to dual way is critical since the boundaries where new information is created consist of coexisting tendencies of unification and diversification.

Since complementary descriptions are intrinsically irreducible and in some cases either contradictory, the conceptual integration or coordination of complementary descriptions requires a higher level description, a new dimension that emerges as a new hierarchical level of complexity, in this case 4D.

The central frame, labeled as the "Self" should describe the interaction of the two ways in duality relation. These ways correspond to construction or forward way, from S to K3, or to deconstruction or backward way, from K3′ to S′.

Figure 1.3 outlines the need of reversing the trend of increasing levels, that is, the front face of the polytope connected to the "Self", by mapping it to the back face of the polytope and the corresponding decreasing order of levels.

This polytopic architecture is proposed as a basic guide, for understanding and solving a large variety of problems, for designing and building emerging smart systems. Further ontological analysis of the systems may require construction of several sub-levels for each level of the general structure.

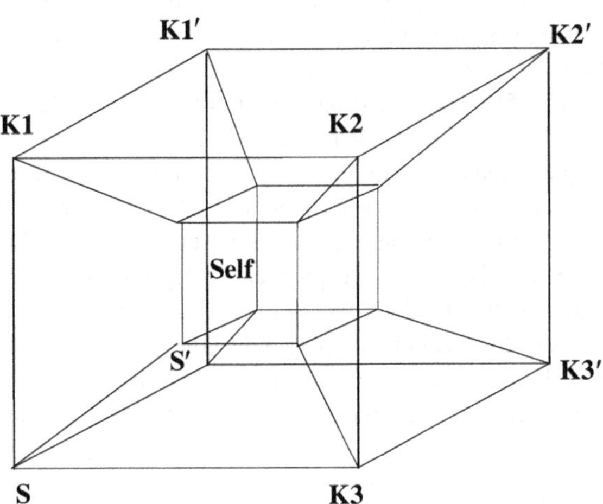

Fig. 1.3 Detailed presentation of the 4D cube

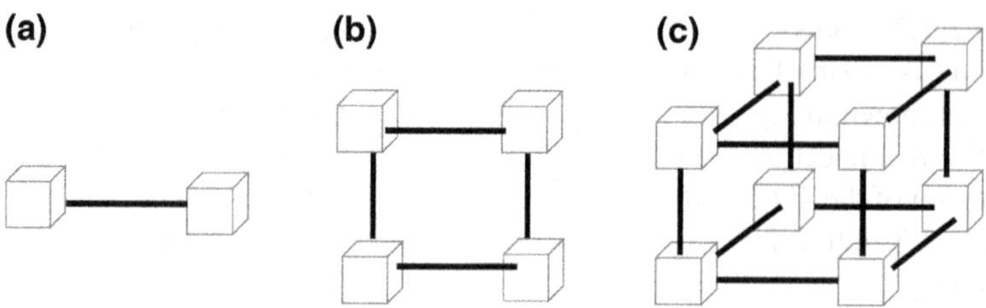

Fig. 1.4 Concept of time

As shown in Fig. 1.1 increasing dimension can be considered as a doubling followed by relating procedure.

The architecture evolves from one cube to two cubes, then to four cubes, then to eight cubes and so on. The reverse process of dimension decreasing is a contracting one.

The doubling procedure towards higher dimensions complements the contracting procedure towards lower dimensions. It should be observed that centers mediate the doubling that is the dimensional increasing.

In engineering design the doubling and contracting correspond to numbering-up and numbering-down procedures (Coppens 2005; Kashid et al. 2010). This general procedure of doubling and contracting illustrates the underlying common ground between methodologies developed for different levels of reality.

The lower dimensions, 0D, 1D, 2D and 3D are well known and close related to space concept in classical geometry and physics. More difficult is to understand 4D, 5D, and 6D frames. For some modern physics researchers, high-dimensionality, that is, the new dimensions are rather time related then space related (Sparling 2007).

Figure 1.4 illustrates possible interpretations of the concept of time for higher dimensions.

The 4D cases, as shown in Fig. 1.4a, may be associated to the linear time. For instance the 4D printing refers to printing of 3D objects that varies in time (Tibbits 2014). It is the linear time shown by the thick hyper-line in Fig. 1.4a, that is the "line of cubes".

The 5D cases, will add a supplementary dimension for time. The planar time may be a useful physical interpretation. Figure 1.4b outlines a cycle of hyper-lines in plane that is a "square of cubes". The 5D object refers to a 3D object that varies in planar time. Cyclic time is an example of planar time.

The 6D situations add another supplementary dimension for time. The spatial time may be a possible physical interpretation. Figure 1.4c outlines a new cycle, and a "cube of cubes" resulting as a parallel universe model. The front and the back faces of the 6D cube correspond to the two parallel universes. The 6D object refers to a 3D object that varies in spatial time. Helical gears time is an example of spatial time. Multi-cyclic time allows modeling the transition between parallel universes.

References

Abe, E.: Hopf Algebras. Cambridge University Press, Cambridge (1980)

Ammon, J.: Hypercube Connectivity within ccNUMA Architectures. SGI Origin Team, Mountain View (1998)

Atiyah, M.F.: Duality in Mathematics and Physics. Conferences FME. University of Barcelona, Barcelona (2008)

Baars, B.J.: The conscious access hypothesis: origins and recent evidence. Trends Cogn. Sci. **6**, 47–52 (2002)

Baas, N.A.: Hyperstructures, topology and datasets. Axiomathes **19**, 281–295 (2009)

Baas, N.A.: Higher order architecture of collections of objects. Int. J. Gen. Syst. **44**(1), 55–75 (2015a)

Baas, N.A.: On higher structures. Int. J. Gen. Syst. **20**, 1–16 (2015b)

Baas, N.A., Ehresmann, A.C., Vanbremeersch, J.P.: Hyperstructures and memory evolutive systems. Int. J. Gen. Syst. **33**(5), 553–568 (2004)

Bainbridge, W.S., Roco, M.C.: Managing Nano-Bio-Info-Cogno Innovations: Converging Technologies in Society. Springer, Berlin (2006)

Barbieri, M.: The Organic Codes: An Introduction to Semantic Biology. Cambridge University Press, Cambridge (2003)

Bertschinger, N., Olbrich, E., Ay, N., Jost, J.: Information and closure in systems theory. In: Artmann, S., Dittrich, P. (eds), Explorations in the Complexity of Possible Life. Proceedings of the 7th German Workshop of Artificial Life, pp 9–21, Amsterdam. IOS Press BV Amsterdam (2006)

Blute, R.: Hopf algebras and linear logic. Math. Struct. Comput. Sci. **6**, 189–217 (1996)

Blute, R., Ehrhard, T., Tasson, C.: A Convenient Differential Category. arXiv preprint. arXiv:1006.3140 (2010)

Blute, R., Scott, P.: The shuffle Hopf algebra and noncommutative full completeness. J. Symbolic Logic **63**, 1413–1435 (1998)

Bonchev, D., Rouvray D.H. (eds).: Complexity in chemistry, biology, and ecology. In: Mathematical and Computational Chemistry, Springer, New York (2005)

Brier, S.: Cybersemiotics: Why Information Is not Enough. University of Toronto Press, Toronto, Canada (2008)

Brier, S.: Cybersemiotic pragmaticism and constructivism. Constructivist Found. **5**(1), 19–39 (2009)

Brown, R., Glazebrook, J.F., Baianu, I.C.: A conceptual construction of complexity levels theory in spacetime categorical ontology: non-abelian algebraic topology, many-valued logics and dynamic systems. Axiomathes **17**(3–4), 409–493 (2007)

Brueckner, S., Di Marzo Serugendo, G., Hales, D., Zambonelli, F. (eds).: Engineering Self-Organising Systems. LNAI 3910, Springer, Berlin (2006)

Buchli, J., Santini, C.: Complexity engineering, harnessing emergent phenomena as opportunities for engineering. Tech. Rep. Santa Fe Institute Complex Systems Summer School, New Mexico, USA (2005)

Ceulemans, E., Van Mechelen, I., Leenen, I.: Tucker3 hierarchical classes analysis. Psychometrika **68**, 413–433 (2003)

Choudhury, N.: World Wide Web and its journey from web 1.0 to web 4.0. Int. J. Comput. Sc. Inf. Tech. **5**(6), 8096–8100 (2014)

Cockett, R.: What is a good process semantics? In: International Conference on Mathematics of Program Construction, pp. 1–3, Springer, Berlin (2006)

Cook, N.D.: The Brain Code: Mechanisms of Information Transfer and the Role of the Center. Methuen, New York (1986)

Coppens, M.O.: Nature inspired chemical engineering learning from the fractal geometry of nature in sustainable chemical engineering. Fractal Geom. Appl.: Jubilee Benoit Mandelbrot **72**, 7–32 (2004)

Coppens, M.-O.: Scaling-up and-down in a nature-inspired way. Ind. Eng. Chem. Res. **44**(14), 5011–5019 (2005)

Coxeter, H.S.M.: Regular Polytopes, 3rd edn. Dover, New York (1973)

Dăscălescu, S., Năstăsescu, C., Raianu, Ş.: Hopf Algebras. An introduction, Pure and Applied Mathematics, **235** Marcel Dekker (2001)

Dehaene, S., Sergent, C., Changeux, J.P.: A neuronal network model linking subjective reports and objective physiological data during conscious perception. Proc. Natl. Acad. Sci. USA **100**, 8520–8525 (2003)

Deleuze, G.: Foucault. University of Minnesota Press, Minneapolis (1988)

Deleuze, G.: The Fold: Leibniz and the Baroque. University of Minnesota Press, Minneapolis (1993)

Edelman, G.M.: Naturalizing consciousness: a theoretical framework. Proc. Natl. Acad. Sci. **100** (9), 5520–5524 (2003)

Ehresmann, A.C.: MENS, an info-computational model for (neuro-)cognitive systems capable of creativity. Entropy **14**(9), 1703–1716 (2012)

Ehresmann, A.C., Vanbremeersch, J.-P.: Memory Evolutive Systems: Hierarchy Emergence, Cognition. Elsevier, Amsterdam (2007)

Ehrhard, T., Regnier, L.: Differential interaction nets. Theoret. Comput. Sci. **364**(2), 166–195 (2006)

Eigen, M., Schuster, P.: The Hypercycle. Springer, Berlin (1979)

Evans, J.S.B.T.: Dual-processing accounts of reasoning, judgment, and social cognition. Annu. Rev. Psychol. **59**, 255–278 (2008)

Fomin, S.: Duality of graded graphs. J. Algebraic Comb. **3**, 357–404 (1994)

Fontana, W.: Algorithmic chemistry: a model for functional self-organization. Artif. Life **II**, 159–202 (1991)

Faulon, J.-L., Bender, A. (eds.): Handbook of Cheminformatics Algorithms. CRC Press, Boca Raton (2010)

Gadamer, H.G.: Truth and Method, 2nd edn. Sheed and Ward Stagbooks, London (2004)

Ganter, B., Wille, R.: Formal Concept Analysis. Mathematical Foundations. Springer, Berlin (1999)

Gilpin, K., Rus, D.: Modular robot systems: from self-assembly to self-disassembly. IEEE Robot. Autom. Mag. **17**(3), 38–53 (2010)

Gilpin, K., Rus D.: A distributed algorithm for 2D shape duplication with smart pebble robots. In: Robotics and Automation (ICRA), IEEE International Conference, pp. 3285–3292, IEEE (2012)

Grossberg, S.: The complementary brain: a unifying view of brain specialization and modularity. Trends Cogn. Sci. **4**, 233–246 (2000)

Haken, H.: Information and Self-Organization: A Macroscopic Approach to Complex Systems, Springer Series in Synergetics, 2nd ed. Springer, Berlin (2000)

Hartmann, N.: The New Ways of Ontology. Greenwood Press, Westport (1952)

Heidegger, M.: Being and Time. Harper and Row, New York (1962)

Iordache, O.: Polystochastic Models for Complexity. Springer, Berlin (2010)

Iordache, O.: Modeling Multi-Level Systems. Springer, Berlin (2011)

Iordache, O.: Self-evolvable Systems. Machine Learning in Social Media. Springer, Berlin (2012)

Iordache, O.: Polytope Projects. Taylor & Francis CRC Press, Boca Raton (2013)

Iordache, O.: Implementing Polytope Projects for Smart Systems. Springer, Cham (2017)

Ji, S.: Complementarism: a biology-based philosophical framework to integrate western science and eastern tao. In: Proceeding of the 16th International Congress of Psychoterapy, pp. 518–548 (1995)

Joni, S.A., Rota, G.C.: Coalgebras and bialgebras in combinatorics. Stud. Appl. Math. **61**(2), 93–139 (1979)

Kaku, M.: Visions: How Science Will Revolutionize the 21st Century and Beyond. Oxford University Press, New York (1999)

Kashid, M.N., Gupta, A., Renken, A., Kiwi-Minsker, L.: Numbering-up and mass transfer studies of liquid–liquid two-phase microstructured reactors. Chem. Eng. J. **158**(2), 233–240 (2010)
Kauffman, S.: The Origins of Order. Self-organization and Selection in Evolution. Oxford University Press, New York (1993)
Kauffman, S.: Investigations. Oxford University Press, New York (2000)
Kelso, J.: The complementary nature of coordination dynamics: self-organization and the origins of agency. J. Nonlinear Phenom. Complex Syst. **5**, 364–371 (2002)
Kelso, J.A.S., Engstrom, D.A.: The Complementary Nature. MIT Press, Cambridge (2006)
Kelso, J.A.S., Tognoli, E.: Toward a Complementary Neuroscience: Metastable Coordination Dynamics of the Brain. In: Murphy, N., Ellis, G.F.R., O'Connor, T. (eds.) Downward Causation and the Neurobiology of Free Will, pp. 103–124. Springer, Heidelberg (2009)
Kleineberg, M.: Integrative Levels. Knowl. Organ. **44**(5), 349–379 (2017)
Langley, P.: Cognitive architectures and general intelligent systems. AI Mag. **27**, 33–44 (2006)
Langley, P., Simon, H., Bradshaw, G., Zytkow, J.: Scientific Discovery: Computational Explorations of the Creative Processes. MIT Press, Cambridge (1987)
Latour, B.: We Have Never Been Modern. Harvard University Press, Cambridge (1993)
Lehn, J.-M.: Dynamic combinatorial and virtual combinatorial libraries. Eur. J. Chem. **5**(9), 2455–2463 (1999)
Lehn, J.-M.: Toward self-organisation and complex matter. Science **295**(29), 2400–2402 (2002)
Lehn, J.-M.: From supramolecular chemistry towards constitutional dynamic chemistry and adaptive chemistry. Chem. Soc. Rev. **36**, 151–160 (2007)
Longo, G., Montévil, M., Kauffman, S.: No Entailing Laws, but Enablement in the Evolution of the Biosphere: arXiv 1201.2069 (2012)
Luhmann, N.: Social Systems. Stanford University Press, California (1995)
Lupasco, S.: Logique et contradiction. Presses Universitaires de France, Paris (1947)
Mandelbrot, B.: The Fractal Geometry of Nature. W.H. Freeman, San Francisco (1982)
Maturana, H., Varela, E.: Autopoiesis and Cognition. The Realization of the Living. Reidel, Dordrecht (1980)
Mossio, M., Longo, G., Stewart, J.: An expression of closure to efficient causation in terms of λ-calculus. J. Theor. Biol. **257**, 498 (2009)
Nicolescu, B.: Manifesto of Transdisciplinarity. SUNY, Albany (2002)
Nicolescu, B.: Methodology of transdisciplinarity: levels of reality, logic of the included middle and complexity. Transdisciplinary J. Eng. Sci. **1**, 19–38 (2010)
Nicolis, G., Prigogine, I.: Self-organization in Nonequilibrium Systems. Wiley, New York (1977)
OMG.: Object Management Group, Software & Systems Process Engineering Meta-Model Specification 2.0 (2008)
Overton, W.F.: Understanding, explanation, and reductionism: Finding a cure for Cartesian anxiety. In: Reductionism and the development of knowledge (pp. 39–62), Psychology Press (2003)
Overton, W.F.: Developmental psychology: philosophy, concepts, methodology. In: Lerner, R.M. (ed).: Handbook of Child Psychology. Vol. 1: Theoretical Models of Human Development, 6th ed., pp. 18–88. Wiley, Hoboken (2006)
Overton, W.F.: Relationism and relational developmental systems: a paradigm for developmental science in the post-Cartesian era. Adv. Child Dev. Behav. **44**, 21–64 (2013)
Pattee, H.H.: Evolving self-reference: matter, symbols, and semantic closure. Commun. Cogn. Artif. Intell. **12**(1–2), 9–28 (1995)
Pattee, H.H.: The physics of symbols: bridging the epistemic cut. Biosystems **60**, 5–21 (2001). (Special Issue: Reflections on the work of Howard Pattee)
Peirce, C.S.: Collected Papers of Charles Sanders Peirce, vol. 1–8. Cambridge University Press, Cambridge (1956)
Piaget, J.: Genetic Epistemology. Columbia University Press, New York (1970)
Piaget, J.: The Construction of Reality in the Child. Ballantine Books, New York (1971)
Piaget, J., Garcia, R.: Psychogenesis and the History of Science. Columbia University Press, New York (1989)

Poli, R.: The basic problem of the theory of levels of reality. Axiomathes **12**, 261–283 (2001)

Poli, R.: Three obstructions: forms of causation, chronotopoids, and levels of reality. Axiomathes **17**, 1–18 (2007)

Pross, A.: Stability in chemistry and biology: life as a kinetic state of matter. Pure Appl. Chem. **77** (11), 1905–1921 (2005)

Ricoeur, P.: Time and Narrative, vol. 1. University of Chicago Press, Chicago (1990)

Ricoeur, P.: From Text to Action. Northwestern University Press, Evanston (1996)

Rocha, L. M.: (ed.) The physics and evolution of symbols and codes: reflection on the work of Howard Pattee. Biosystems **60**(1–3) (2001)

Rosen, R.: Life itself: a comprehensive enquiry into the nature, origin and fabrication of life. Columbia University Press, New York (1991)

Schwab, K.: The Fourth Industrial Revolution. World Economic Forum, Geneva (2016)

Seitz, C.L.: The cosmic cube. Commun. ACM **28**(1), 22–33 (1985)

Sharp, P.A., Langer, R.: Promoting convergence in biomedical science. Science **333**, 527 (2011)

Sparling, G.A.J.: Germ of a synthesis: space-time is spinorial, extra dimensions are time-like. Proc. R. Soc. Lond. Ser. A Math. Phys. Eng. Sci. **463**, 1665–1679 (2007)

Sweedler, M.E.: Hopf Algebras, Mathematics Lecture Note Series. W. A. Benjamin Inc., New York (1969)

Thiel, T.: The design of the connection machine. Des. Issues **10**(1), 5–18 (1994)

Tibbits, S.: 4D printing: multi-material shape change. Architectural Des. **84**(1), 116–121 (2014)

Van Den Heuvel, M.P., Sporns, O.: Rich-club organization of the human connectome. J. Neurosci. **31**(44), 15775–15786 (2011)

Velardo, V.: Recursive ontology: a systemic theory of reality. Axiomathes **26**(1), 89–114 (2016)

von Uexküll, J.: Theoretische Biologie..: Suhrkamp Taschenbuch Wissenschaft Frankfurt a. M (1973)

Whitehead, A.N.: Process and Reality: An Essay in Cosmology. Free Press, New York (1978)

Ziegler, G.M.: Lectures on Polytopes. Graduate Texts in Mathematics, vol. 152. Springer, New York (1995)

Yim, M., Shen, W.M., Salemi, B., Rus, D., Moll, M., Lipson, H., Klavins, E., Chirikjian, G.S.: Modular self-reconfigurable robot systems [grand challenges of robotics]. IEEE Robot. Autom. Mag. **14**(1), 43–52 (2007)

Chapter 2
Integration and Separation

2.1 Tree Structure

2.1.1 Rooted Trees

Rooted trees are examples of combinatorial objects which appear in different contexts in science and industry. They describe multi-scale systems, taxonomy, separation schemes, schedules, automata self-reconfiguration, logical schemes, organizational structures, and so on. Combinatorial Hopf algebras emerged as the appropriate instrument in the study of composition and decomposition processes. The product or multiplication, denoted by ∇ or $\bigcirc$, describes the construction while the coproduct or comultiplication, denoted by Δ, describes the deconstruction processes (Appendix B) (Dăscălescu et al. 2001).

Hopf algebras have been associated to different types of trees.

Could be mentioned here the Hopf algebras of Connes and Kreimer (1998, 2000), the Hopf algebras of Grossman and Larson (1989, 1990), the dendriform Hopf algebras of Loday and Ronco (1998), the Hopf algebra introduced by Brouder and Frabetti (2003), and the Hopf algebras studied by Hoffman (2003, 2008).

A rooted tree is a tree with a distinguished vertex (which we may place at the bottom).

A forest is a string of rooted trees. The weight of a tree is the number of vertices, and the weight of a forest is the sum of the weights of its constituent trees. The smallest tree is denoted by • and consists of just the root.

We write $\varnothing$ for the empty forest.

There is a map B_+ that takes any forest to a tree of weight one greater, by connecting them all via a new root. Figure 2.1 shows the map B_+.

For a distillation column the root could denote the reboiler while other dots correspond to separated compounds from a mixture.

B_+ has an inverse B_- that cuts off the root, leaving the forest of its descendents.

O. Iordache, *Advanced Polytopic Projects*, Lecture Notes in Intelligent Transportation and Infrastructure,
https://doi.org/10.1007/978-3-030-01243-4_2

Fig. 2.1 The map B_+

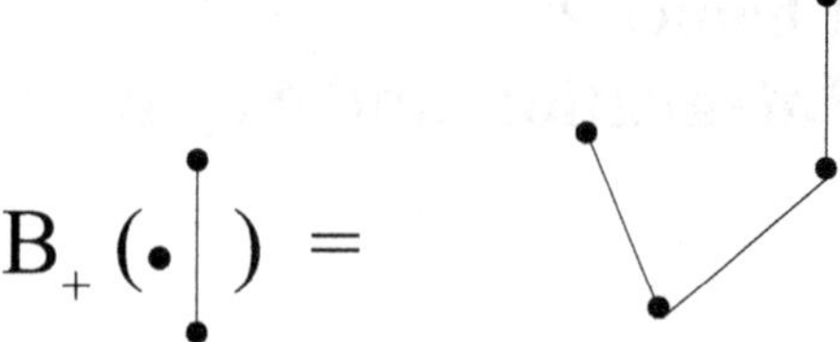

Denote by kT the Hopf algebra of rooted trees studied by Grossmann and Larson (1989).

Let kT be the vector space generated by rooted trees and graded by the number of non-root vertices. There is a multiplication on kT defined by Grossman and Larson. For rooted trees t, t′, let $B_-(t) = t_1 \ldots t_n$, and let m be the number of vertices of t′. Then the product denoted by t ○ t′ is the sum of the m^n trees formed by attaching the root of each t_i to a vertex of t′. If t = • we define t ○ t′ to be t′.

Observe that: t○• = $B_+(B_-(t))$ = t, so • is a two-sided identity.

This multiplication is associative, but not commutative.

Figure 2.2 illustrates the product for kT.

The product describes the construction of the separating device or separation scheme.

The coproduct on kT is given by Eq. (2.1):

$$\Delta(t) = \sum_{\alpha \cup \beta = \{1,\ldots,k\}} B_+(t_\alpha) \otimes B_+(t_\beta) \tag{2.1}$$

Here $B_-(t) = t_1 \ldots t_k$, the sum is over all partitions of {1, …, k} into two subsets, and t_α is the forest consisting of all t_i, $i \in \alpha$.

The coproduct describes the deconstruction of the separation scheme.

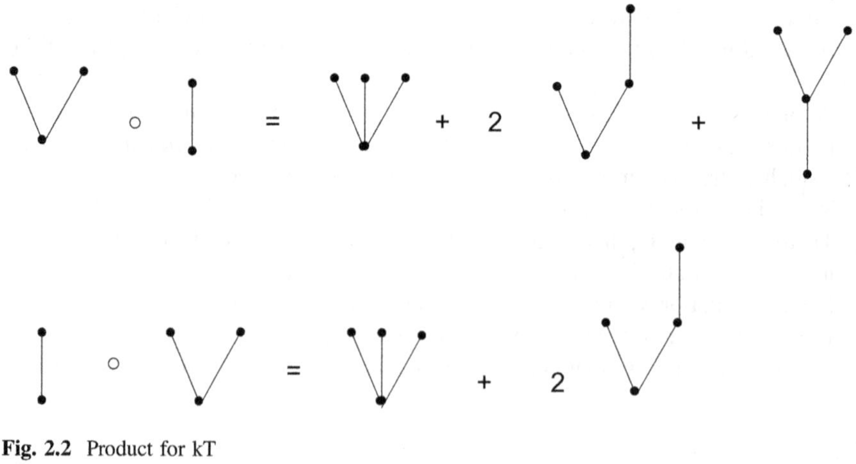

Fig. 2.2 Product for kT

Just coproducts and products generate any separation and integration scheme. This explains the utility of Hopf algebras.

Primitive trees are those rooted trees t such that $B_{-}(t)$ consists of a single tree: in fact all primitives of kT are linear combinations of primitive trees.

The Connes-Kreimer Hopf algebra is the commutative algebra of forests of rooted tree, with product given by juxtaposition and coproduct Δ given by requiring that Δ is multiplicative and that for all root tree t:

$$\Delta(t) = t \otimes 1 + (id \otimes B_{+})(\Delta(B_{-}(t)) \tag{2.2}$$

The Connes-Kreimer Hopf algebra is the dual of the Grossman-Larson Hopf algebra kT of trees.

Let kP denote the Hopf algebra of planar rooted trees (Hoffman 2008).

We can make the vector space kP generated by the planar rooted trees into an algebra as follows. To form t O t′ with $B_{-}(t) = t_1, t_2, \ldots, t_n$, attach the roots of each t_i in order to the vertices of t′, respecting the natural order of the vertices of t′. We put t O t′ = t′ if t = •.

Figure 2.3 illustrates the product for kP.

The coproduct on kP looks similar but not identical to that on kT.

If $B_{-}(T) = T_1, T_2, \ldots, T_k$, then

$$\Delta(t) = \sum_{i=0}^{k} B_{+}(T_1, \ldots, T_i) \otimes B_{+}(T_{i+1}, \ldots, T_k) \tag{2.3}$$

Note that this coproduct is not co-commutative.

Knowledge of the associated Hopf algebra allows automatic design by construction and deconstruction of the separation schemes.

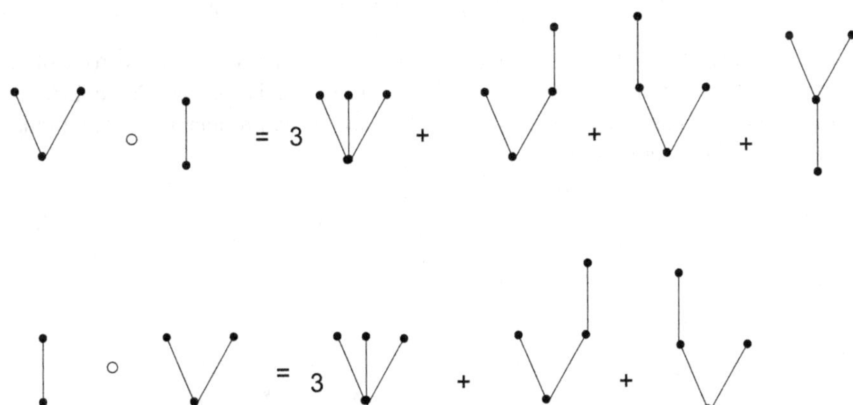

Fig. 2.3 Product for kP

The selection of Hopf algebra is based on physical and engineering knowledge allowing reasonable choice out of mathematical possibilities.

This refers to the conditions of joining two separation devices as specified by product or to breaking a device or a scheme in component devices as specified by coproduct.

2.1.2 Ordered Trees

Ordered trees with 1, 2, 3, and 4 nodes are shown In Fig. 2.4. (Stanley 1997; Aguiar and Sotille 2005):

Given two ordered trees x and y, we may join them together at their roots to obtain another ordered tree x\y, where the nodes of x are to the left of those of y.

An ordered tree is planted if its root has a unique child. Every ordered tree x has a unique decomposition: $x = x_1\backslash\ldots\backslash x_k$ into planted trees $x_1, \ldots, x_k$, corresponding to the branches at the root of x.

These are the planted components of x.

The set of nodes of an ordered tree x is denoted by Nod(x). Let x be an ordered tree and $x_1, \ldots, x_k$, its planted components, listed from left to right and possibly with multiplicities.

Given a function f: $[k] \rightarrow Nod(y)$ from the set $[k] = \{1, \ldots, k\}$ to the set of nodes of another ordered tree y, it is possible to form a new ordered tree $x \neq_f y$ by identifying the root of each component x_i of x with the corresponding node f(i) of y. For this to be an ordered tree, retain the order of any components of x attached to the same node of y, and place them to the left of any children of that node in y. Given a subset $S \subseteq [k]$, say

$$S = \{i_1 < \cdots < i_p\} \text{ and let } x_S := xi_1\backslash\ldots\backslash xi_p$$

Equivalently x_S is the tree obtained by erasing the branches at the root of x which are not indexed by S. Let $S^C = [k]\backslash S$. Here C stands for complementary.

Equations (2.4) and (2.5) defines the product and the coproduct associated to the Hopf algebra of ordered trees.

$$x \circ y = \sum_{f:[k]\rightarrow Nod(y)} x \neq_f y \tag{2.4}$$

Fig. 2.4 Ordered trees

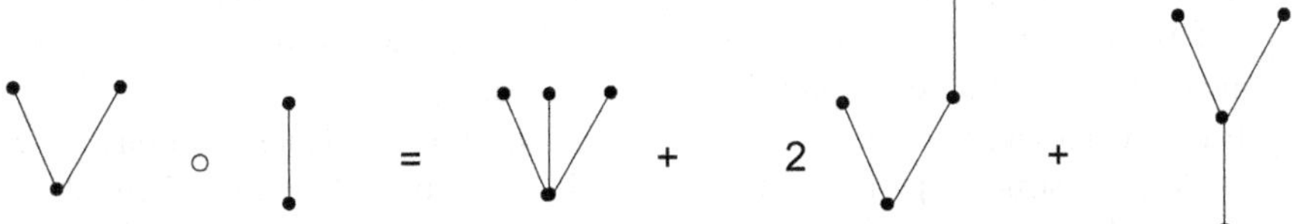

Fig. 2.5 Product for ordered trees

$$\Delta(x) = \sum_{S \subseteq [k]} x_S \otimes x_{S^C} \tag{2.5}$$

Figure 2.5 shows an example of product for ordered trees.

A heap ordered tree is an ordered tree x together with a labeling of the nodes, a bijection

Nod (x) → {0, 1, …, n} such that:

- The root of x is labeled by 0
- The labels increase as we move from a node to any of its children
- The labels decreases as we move from left to right within the children of each node.

The heap ordered trees with 1, 2, 3 and 4 nodes are shown in Fig. 2.6.

The difference between the trees shown in Figs. 2.5 and 2.6 is due to labeling.

Let x and y be heap ordered trees. Suppose x has k planted components.

Given a function f: [k] → Nod(y) the ordered tree $x \neq_f y$ may be turned into a heap ordered tree by keeping the labels of y and implementing the labels of x uniformly by the highest label of y. Given a subset S ⊆ [k], S = $\{i_1 < \cdots < i_p\}$ the ordered tree x_S may be turned into a heap ordered tree by standardizing the labels, which is to replace the ith smallest label by the number i for each i (Aguiar and Sotille 2005).

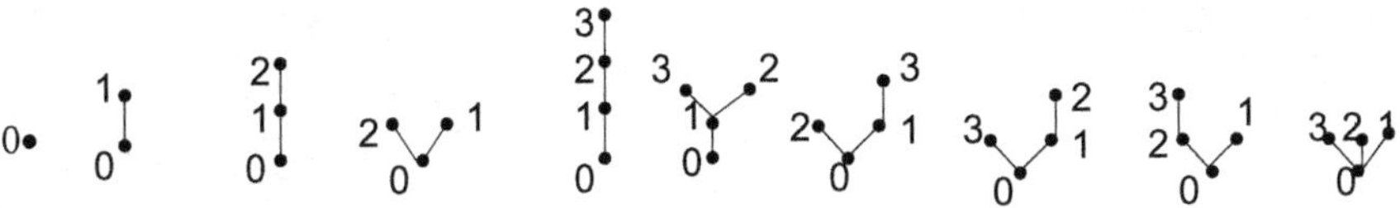

Fig. 2.6 Heap ordered trees

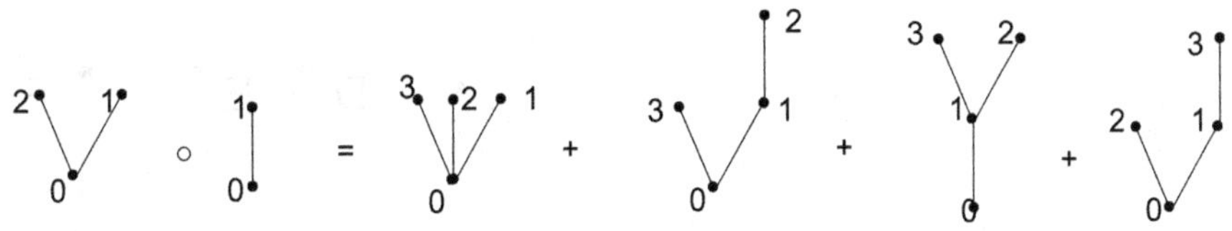

Fig. 2.7 Product for heap ordered trees

Given heap ordered trees and using the product and the coproduct defined by Eqs. (2.4) and (2.5) it is possible to define the Hopf algebra of heap ordered trees.

Figure 2.7 shows an example of product for heap ordered trees.

Heap ordered trees on n + 1 nodes are in bijection with permutations on n letters. We construct a permutation from such a tree by listing the labels of all non-root nodes in such way that the label of a node i is listed to the left of the label of a node j precisely when i is below or to the left of j (that is, when i is a predecessor of j, or i is among the left descendants of the nearest common predecessor between i and j). For instance, the six heap ordered trees on 4 nodes shown in Fig. 2.6 correspond respectively to 123, 132, 213, 312, 231, and 321.

Heap ordered trees are useful tool to describe organizational structures.

2.1.3 *Organizational Trees*

An attempt to use rooted tree to model human organization is presented in the following (Pang 2016).

A company starts with n_0 employees in a tree structure, where each employee except the manager has exactly one direct superior. The manager or owner, denoted here by man, is the direct superior of the department heads. For example, Fig. 2.8 shows the structure of a company with $n_0 = 8$, employees denoted by A to G.

The rooted tree has a reversed presentation in this case.

The company has two departments: A heads the left department, consisting of himself and B, while C heads the right department, of C, D, E, F and G. C is the direct superior to D, E and F. Firing may be based on performances. Let q be a parameter between 0 and 1. The yearly performance of each employee is, independently, uniformly distributed between 0 and 1, and each month all employees with performance below 1 − q are fired. Each firing causes a set of promotions.

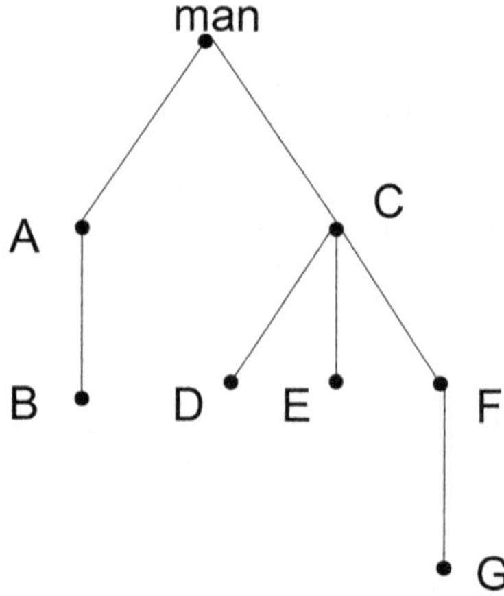

Fig. 2.8 Organization with a tree structure

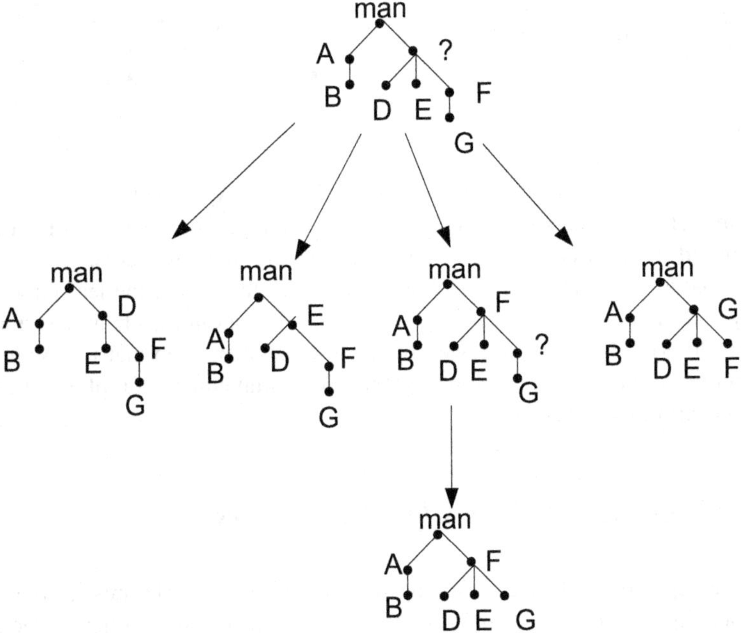

Fig. 2.9 Promotion scenarios

The chain keeps track of the tree structure of the company, but does not see which employee is taking which position.

The situation is as follows: The organization has a tree structure, so each employee except the manager has exactly one direct superior. Each year, some employees are fired and each firing independently causes a cascade of promotions: first, someone further down the chain of superiority from the fired employee is uniformly selected to replace him. Then, if the promoted employee was superior to anyone, then one of those is uniformly selected and promoted to his position. This process continues until someone who is not superior to anyone, a leaf in the tree, is promoted.

The chain keeps track of the tree structure of the company, but does not know which employee is taking which position.

Figure 2.9 shows the promotion scenarios.

A connection to the decorated Connes-Kreimer Hopf algebra of trees may be established (Connes and Kreimer 1998; Pang 2016). The labels or decorations are the job positions.

A basis is the set of all decorated rooted forests—that is, each connected component has a distinguished root vertex, and each non-root vertex is assigned one of a finite set of labels.

The degree of a forest is its number of vertices.

The product of two forests is their disjoint union, preserving all labels. This Hopf algebra is commutative.

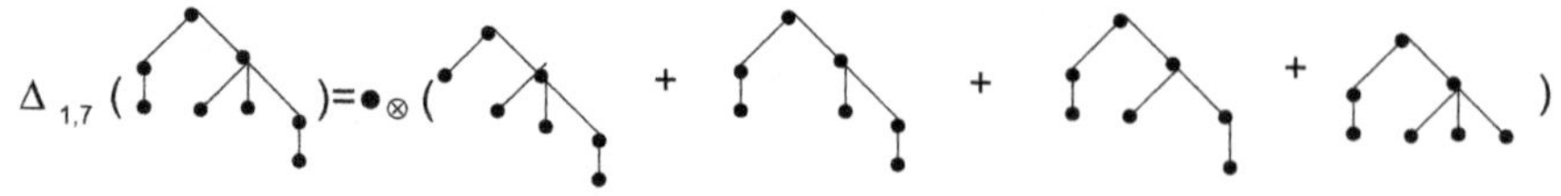

$\Delta_{1,7}$ ( ) = • ⊗ (+ + +)

Fig. 2.10 Partial coproduct

The partial coproduct of a forest x on n vertices is $\Delta_{1,n-1}$ (x):

Figure 2.10 illustrates the partial coproduct (Pang 2016). The tree of 8 dots is decomposed in the fired 1 dot coupled to the resting trees of 7 dots.

The goal is to use the Hopf-algebraic framework to analyze the leaf-removal or employee firing chain. The leaf-removal step also occurs in the chain described by Fulman (2009), where it is followed by a leaf-attachment step that is absent here.

Selection of the associated Hopf algebra allows implementation of an automatic firing and promotion processes.

2.1.4 Dual Graded Graphs for Rooted Trees

There are separations devices allowing the splitting of the mixtures in more than two fractions. Separation columns with multiple feed and product collected at several levels in the column are of this type.

Multiple separations may be described by rooted trees generalizing binary rooted trees.

As defined, any rooted tree is a partially ordered set whose elements are called vertices with a unique lowest element, the root vertex, such that for any vertex, v the vertices exceeding v in the partial order form a chain. If v exceeds w in the partial order, we call w a descendant of v and v an ancestor of w.

Rooted trees do not distinguish between two separations schemes that can be obtained by simple rotation around a vertex in the schemes. This is the case if the compounds are considered as separated no matter if they are in the light, intermediary or heavy phase.

To separate is the primary objective, no matter the mixture and the phase. This corresponds to unlabelled rooted trees.

Rooted trees of this type have been described by Hoffman (2003, 2008) and by Sloss (2005).

A partial order on this set considers that t $\leq$ t′ if and only if t can be obtained by deleting some set of non-root vertices of t′.

Under this definition is clear that t < t′, that is t is covered by t′, if and only if t is obtained by deleting exactly one leaf vertex of t′. Thus the set of rooted trees is graded by the number of non-root vertices. The study makes use of differential posets (Stanley 1988) and dual graded graphs (Fomin 1994) (Appendix A).

Figure 2.11 shows the rooted trees U-graph.

The U-graph details the construction of separation scheme.

Figure 2.12 shows the rooted trees D-graph.

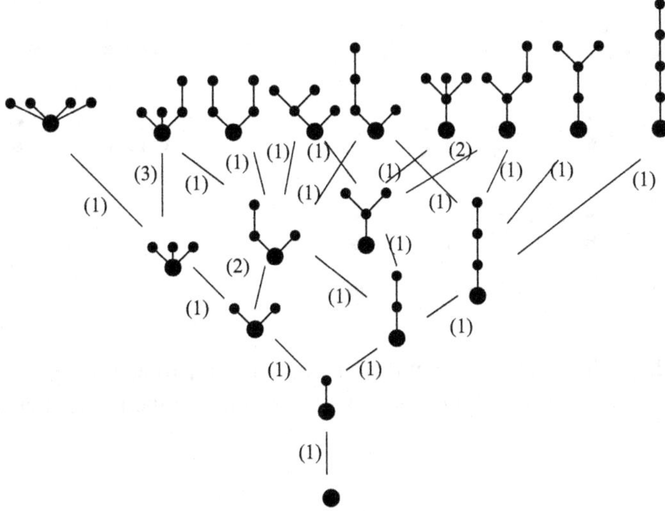

Fig. 2.11 Rooted trees U-graph

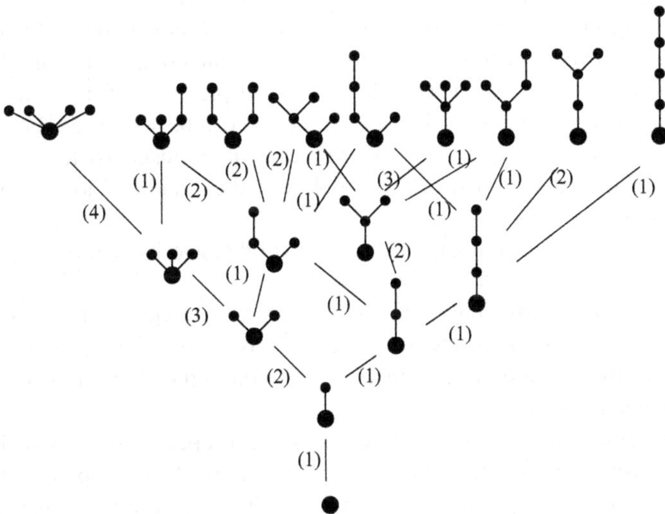

Fig. 2.12 Rooted trees D-graph

The D-graph details the deconstruction of separation scheme.

The weights indicated on the connections between two rooted tree t′ and t show the number of ways we can obtain t′ from t.

For the same connection line in graphs the weights are different in the U-graph and D-graph.

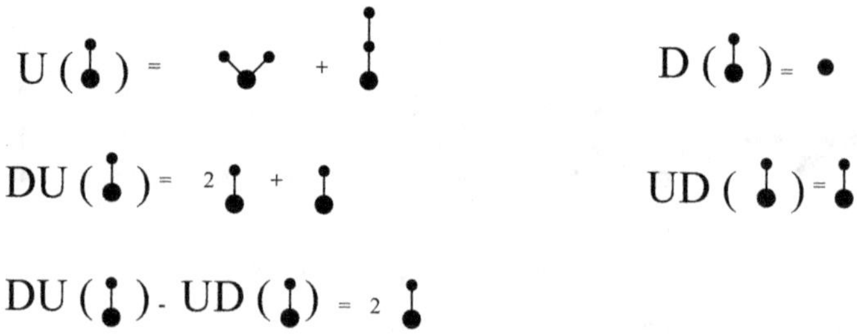

Fig. 2.13 Commutation condition for rooted trees

Figure 2.13 illustrates the commutation condition for rooted trees.
It shows how the differential model works to generate the rooted trees.

2.1.5 *Binary Trees and Hypercubes*

The relation between lifted or developed binary trees and the separation schemes is based on the rules to obtain the dual trees (Iordache 2013, 2017).

Vectors defining the vertices in the trees may be naturally associated to chemical compounds. To every species in a mixture a vector the components of which take only two values "1" or "0" where "1" means the presence of a given structural element or property whereas "0" means its absence, can be associated. For instance, "1" may correspond to high volatility, whereas "0" corresponds to low value of volatility, "1" may correspond to a hydrophilic radical and "0" to the absence of this, and so on.

Vectors associated to different compounds are denoted by: $i = [i_1, i_2, \ldots, i_k, \ldots]$ where i_k are either "1" or "0".

A hierarchy of the structural elements or properties is required. For instance, it is considered that the property indexed by i_1 is more significant that the property indexed by i_2, this more significant that i_3, and so on in the order of the coordinates in the associated vectors.

We may associate several digits to the same property. With two digits we characterize very high, high, low and very low values of the property. The corresponding vectors are: [11], [10], [01] and [00]. With three digits we may associate eight characterizations for that property and so on.

To associate a graph to a configuration we examine the coordinates of the vectors describing the compounds in the order of significance.

We look to first coordinate. If it is the same we put the compounds in the same class. Otherwise we put compounds with "1" to the left and that with "0" to the right in a tree.

Left or right may be associated to light and heavy phases in separation to rectified and stripped phase in a distillation column, to non-adsorbed and adsorbed phase in a chromatograph.

Then we look to the second coordinate and perform the same separation in classes and the vectors evaluation process continues for a finite number of steps.

Initial steps of the developed variants of binary trees are shown in Fig. 2.14.

They have been studied as infinite 2-nary trees (Fomin 1994; Qing 2008). Only 3 steps have been represented here.

The vertex set is represented by strings of "1" or "0" of finite length. The length of the string is the rank of the element.

For the U-graph shown in Fig. 2.14a, x is covered by y iff y can be obtained from x by adding a number to the right end of x. The D operator corresponds to deleting a letter that is a digit in the associated vector.

In the D-graph a word x of rank k is linked with a word y of rank k-1 by a vertex having as weight the number of ways to obtain y from x by deleting a letter from x.

The weights are indicated on the tree shown Fig. 2.14b if different than one.

For instance, Fig. 2.14b shows the weight (2) for the vertices connecting 100 to 10 and 011 to 01.

Figure 2.15 illustrates the commutation condition for developed binary trees.

The elements of different levels in the binary tree may be associated to the binary vectors of the polytopes of different dimensionality.

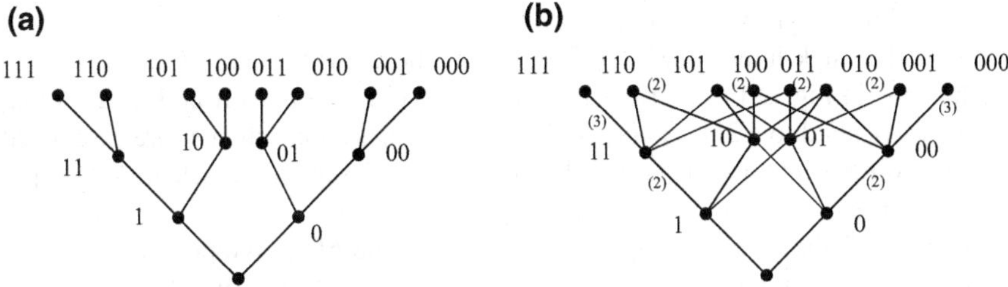

Fig. 2.14 Binary trees

$$U(1) = 11 + 10 \qquad D(1) = \hat{0}$$

$$DU(1) = 2\text{x}1 + 1 + 0 \qquad UD(1) = 1 + 0$$

$$DU(1) - UD(1) = 2\text{x}1$$

Fig. 2.15 Commutation condition for binary trees

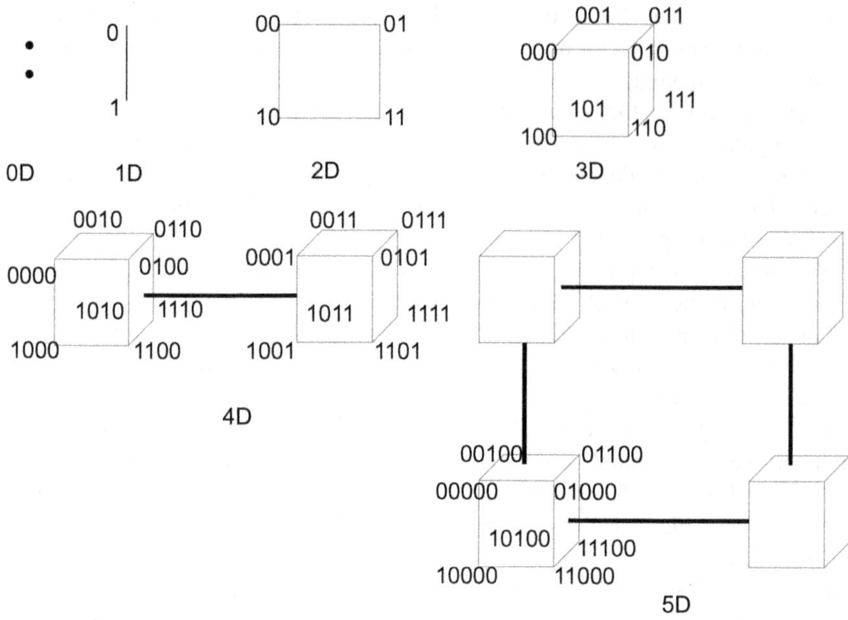

Fig. 2.16 Binary vectors and polytopes

Figure 2.16 illustrates the relation between the binary vectors on trees and the corresponding polytopes from 0D to 5D.

Using the thick hyper-lines, the 5D cube is shown as "square of cubes".

The cube in Fig. 2.16 describes the 3-digits level in the trees from Fig. 2.14 that may be associated to a 3D separation. Innovative separation schemes are associated to higher dimensionality. This may be considered as separation schemes of separation schemes.

Figure 2.17 shows binary vectors associated to the 6D polytope.

With the thick hyper-lines, the 6D cube is shown as "cube of cubes".

A trajectory of a separation or integration scheme may be shown on the U-tree or D-tree but also on the corresponding polytopes.

Figure 2.18 summarizes the relation between binary trees and polytopes.

The doubling procedure from 0D to 6D towards higher dimensions complements the contracting procedure from 6D to 0D towards lower dimensions. Construction complements deconstruction.

High-dimensional schemes have been investigated for parallel computers (Seitz 1985; Ammon 1998). In chemical engineering it is an interest to organize and evaluate parallel and sequential separations.

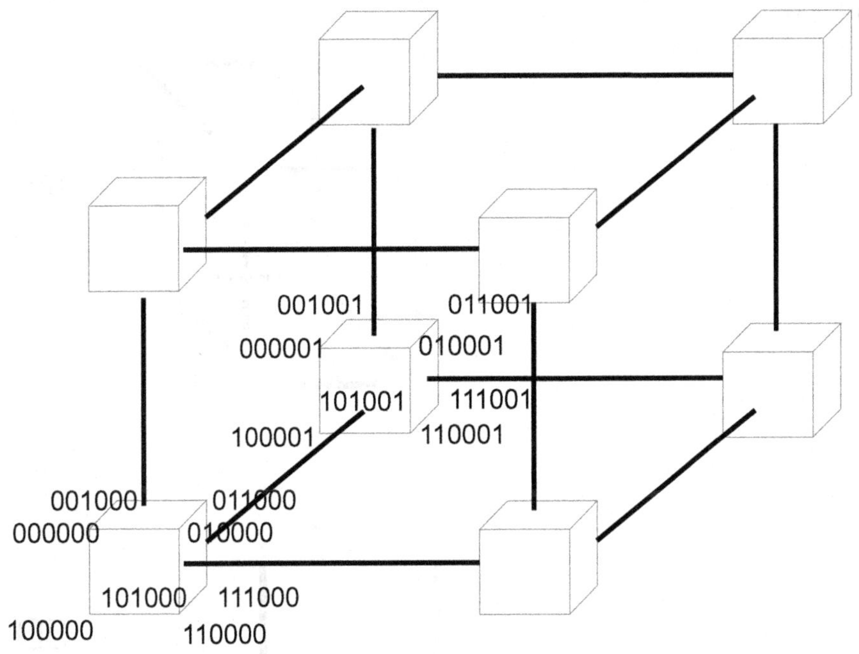

Fig. 2.17 Binary vectors and polytope 6D

2.2 Polytopic Separations

2.2.1 Rooted Tree Polytope

Polytopic separation develops the periodic processes of separation by coupling a cyclic process with the associated reverse cycling processes.

We refer to the coupling of both directions U and D.

The operator U describes the configuration process starting from low number of separated compounds in the scheme. It is the way: S → K1 → K2 → K3.

Here S, K1, K2, K3 corresponds to schemes with increasing number of separated compounds. It is the construction way from small to large.

The operator D describes schemes from higher towards lower number of separated compounds.

It is the deconstruction way: K3′ → K2′ → K′ → S′ from large to small.

Figure 2.19 shows the polytope 4D for rooted trees.

The polytope 4D shown in Fig. 2.19 illustrates the architecture of potentially self-evolvable and smart systems. It is the representation known as hypercube or 4D cube and consisting from a pair of 3D cubes. The 4D cube is obtained by joining all corners of the first 3D cube with the corresponding corners of the second 3D cube.

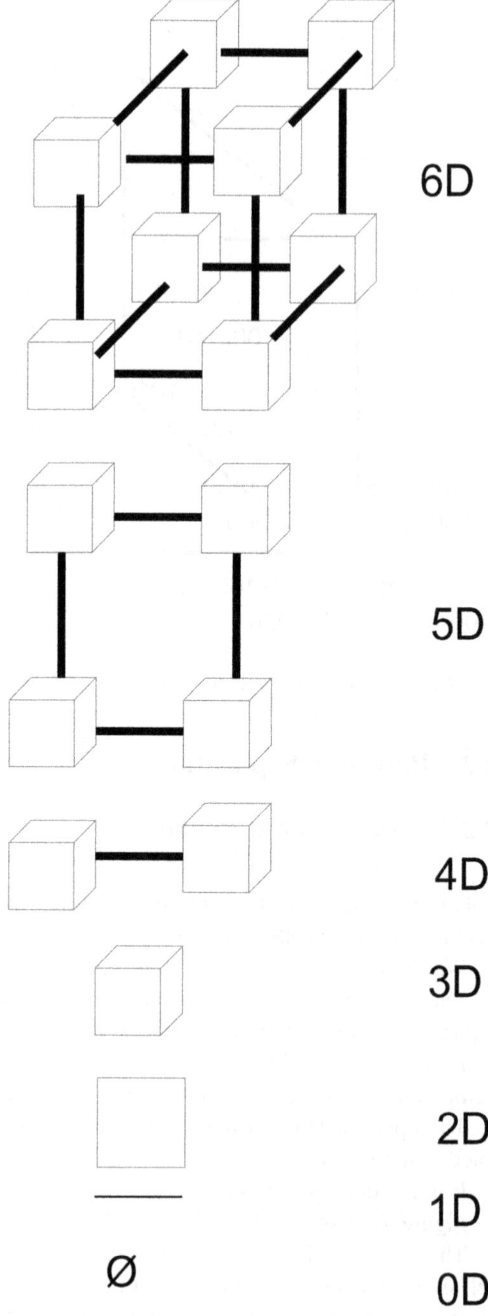

Fig. 2.18 Binary trees represented as polytopes

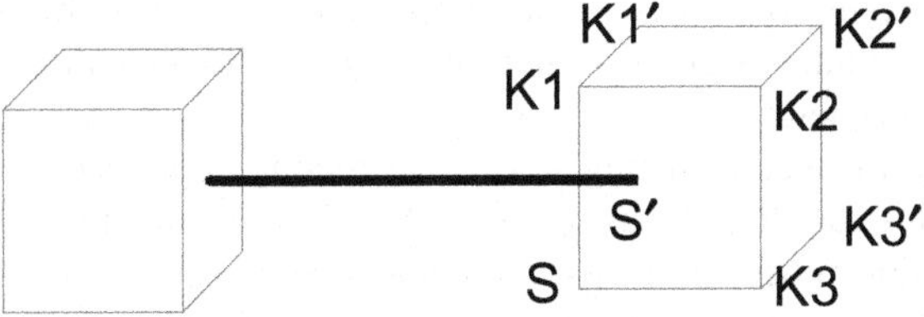

Fig. 2.19 Polytope 4D for rooted trees

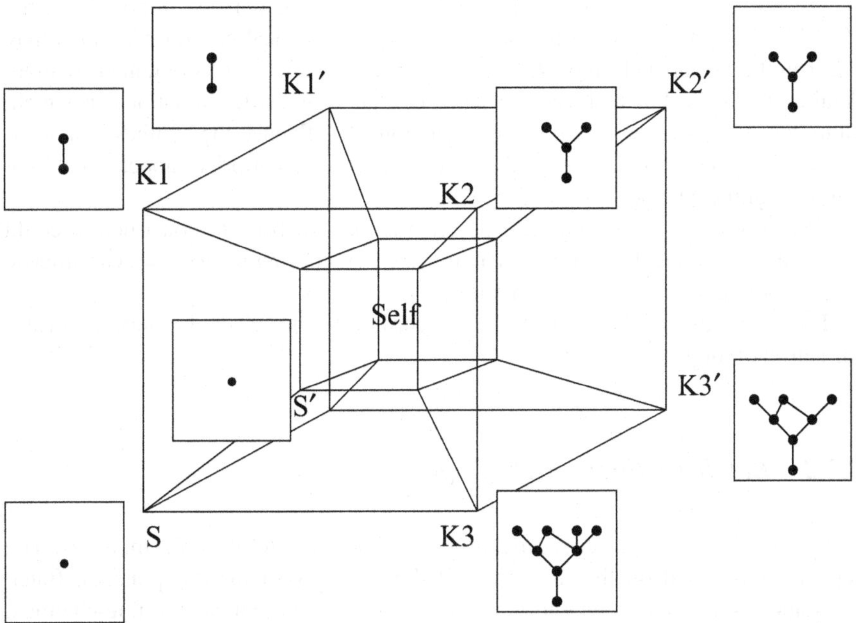

Fig. 2.20 Polytope 4D for rooted trees configurations

The 8 lines connecting the two 3D cubes are represented in Fig. 2.19 by a thick hyper-line. In this way the 4D cube has a simplified representation.

An equivalent representation of a 4D cube is shown in Fig. 2.20. The first cube from Fig. 2.19 is represented as the "Self" inner cube in Fig. 2.20.

Figure 2.20 shows the polytope 4D for rooted trees configurations.

Figure 2.20 illustrates the increasing levels trend way S → K1 → K2 → K3, that is, the front face of the polytope. Connected to the "Self", by mapping it to the back face of the polytope and the corresponding decreasing levels trend is the way: K3′ → K2′ → K′ → S′.

The front face represents the scheme construction, from small to large, while the back face represents the deconstruction of the scheme, from large to small.

The right coupling of separation and integration of dual flow-sheets based on rooted trees would allow significant improvements in the separation if compared to single way approaches.

The swing between constructing and deconstructing steps needs a supplementary dimension, in this case the 4D, as shown in Fig. 2.20. Correlating the constructing and deconstructing processes helps us to make combinations or separations that could not be made or imposed by only one of the dual ways.

Higher dimensions as 5D and 6D dramatically increase the potentialities.

There is an accumulation of high-dimensional architectures waiting to be explored.

The swing between constructing and deconstructing may be imposed via the "Self" by heuristic rules, by thermodynamic or economical criteria. Conflicts may occur in the use of such rules. Heuristics could contradict results obtained by more detailed methods. For instance, the energetic analysis shows situations in which, although the properties of the present compounds influence the synthesis of separation sequences, variable feed and product compositions or prices appear to exhibit a more significant role.

Transition from construction to deconstruction and back to construction could follow the economical or environmental cycles. According to these we can impose criteria as maximum or minimum entropy production.

The "Self" guides the entire system according to the selected criteria and rules for self-evolution.

2.2.2 Batch Distillation Polytope

The distillation is a widespread method for the separation of liquid mixtures. The separation is based on the difference of the volatility of the components. Batch distillation is a frequent separation process in the pharmaceutical and fine-chemical industries.

Figure 2.21 shows batch distillation configurations.

The most frequently used batch distillation configuration is the batch rectifier (BR). Figure 2.21a shows a batch rectifier configuration. In the case of a batch stripper (BS) the vessel ensuring the feed arriving at the top stage of the column is located above the column (Fig. 2.21b).

Nonconventional configurations, as for instance batch stripper (Fig. 2.21b) and middle vessel column (Fig. 2.22) received also attention (Demicoli and Stichlmair 2003; Kotai et al. 2006).

The operation of the different configuration is presented for the separation of a ternary mixture where the order of decreasing volatility is A, B, C. Each configuration needs heating at the bottom (reboiler) and cooling at the top (condenser) of the equipment.

In the case of a batch rectifier (BR) the charge is filled in a heated, great volume vessel (reboiler) which is located under the column (Fig. 2.21a). The major part of

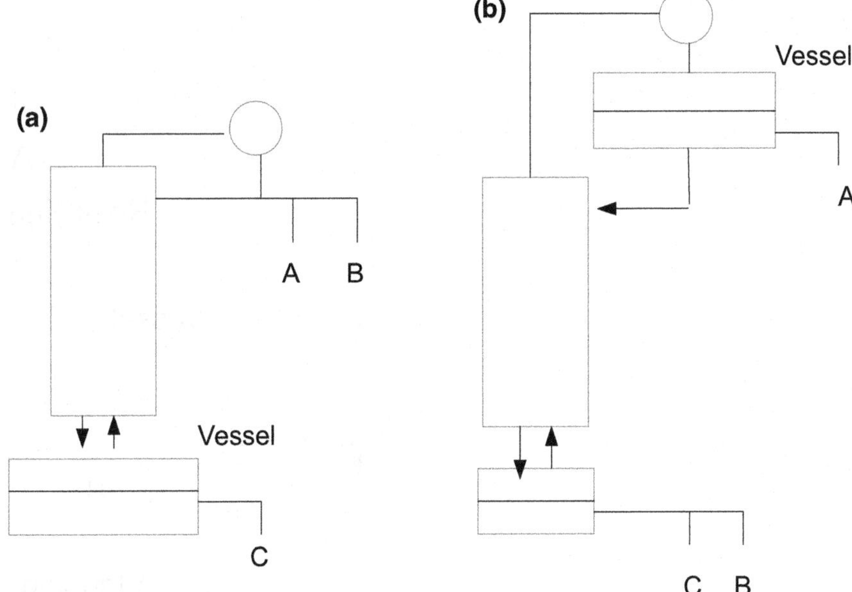

Fig. 2.21 Batch distillation configurations

the separation is performed by the column, equipped with a condenser, where the vapor continuously arriving from the reboiler, having varying composition, is separated to distillate and liquid flowing back to the reboiler. Hence the BR can be considered as a bottom vessel column, as well. In the distillate first A (1st cut) then B (2nd cut) can be obtained as product. C remains in the vessel and can be gained in the residue. Since in the product stream distillate, the more volatile components are enriched, each stage temperature increases with the time.

In the case of a batch stripper (BS) the vessel ensuring the feed arriving at the top stage of the column is located above the column (Fig. 2.21b). Therefore this equipment can be considered as a top vessel column. The charge is filled in this top vessel. The condensate coming from the condenser is also introduced here. The product is continuously withdrawn from the partial reboiler at the bottom of the column. Hence first C (the heaviest component) then B can be obtained as product. The lightest component A remains in the vessel. Since in the product stream bottoms the less volatile components are enriched each stage temperature decreases with the time contrary to the BR.

The middle vessel column (MVC) is a combination of the BR and BS (Fig. 2.22).

Figure 2.22 shows the middle vessel column (Kotai et al. 2006).

The charge is filled in the vessel located at the middle of the column between two column sections. From the middle vessel the liquid of varying composition is fed to the top stage of the lower (stripping) section. Both the liquid leaving the

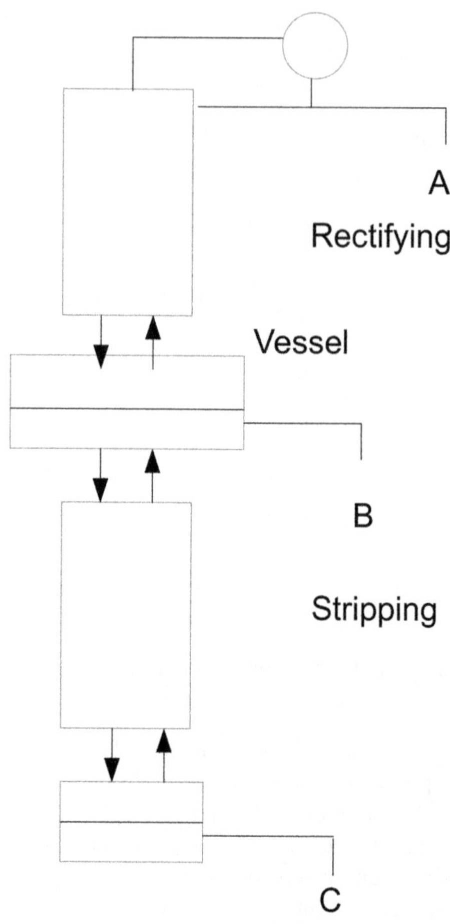

Fig. 2.22 Middle vessel column

bottom stage of the upper (rectifying) section and the vapor leaving the top of the stripping section are introduced into the middle vessel. The vapor leaving the vessel arrives at the bottom of the rectifying section. The vessel can be considered as a stage with great, varying hold-up. Products are continuously withdrawn from both the top (A can be produced in the distillate) and the bottom (C can be produced in the bottoms).

At the end of the operation in the middle vessel B can be enriched.

The middle vessel column may be described by the 4D polytope shown in Fig. 2.20.

We refer to the coupling of both directions rectifying and stripping.

The rectifying operation describes the configuration process starting from lower number of separated compounds towards higher number of separations in the scheme. It is the way: S → K1 → K2 → K3.

Here S, K1, K2, K3 corresponds to schemes with increasing number of separated compounds as the level increases. It is the construction from small to large.

The stripping operation describes schemes from higher towards lower number of separated compounds. It is the way: K3′ → K2′ → K′ → S′, the deconstruction from large to small.

Several strategies have been proposed to increase the operating efficiency of batch distillation (Flodman and Timm 2012). Conventional batch rectification and inverted batch stripping are used cyclically to promote high product flow rates for a binary fractionation. Process controls are implemented to maintain constant product purity specifications by varying the slope of the operating line. While rectifying, the light component is removed as distillate, concentrating the heavy component in the reboiler. As a result, the distillate rate decreases with time. The column is then changed from rectification to stripping modes, and the heavy component is removed as bottoms product, concentrating the light component in the distillate drum. This causes the bottoms rate to diminish with time, and the column is once again converted back to rectifying mode. Cyclic operation, transitioning from batch rectifying to stripping back to rectifying, continues until all of the initial charge is fractionated or is combined with a new charge (Flodman and Timm 2012).

The above described strategy illustrates the swing between constructing and deconstructing steps that implies a supplementary dimension in this case the 4D. Correlating the constructing and deconstructing processes helps us to run separation operations that could not be done by only one of the dual batch distillation operations considered separately.

The process illustrated in Fig. 2.22 is polytopic since it includes both constructing and de-constructing devices and the middle vessel assuring and controlling the 4D polytope processing. MVC appears as a "Self" exemplar. This "Self" guides the entire separation system according to the selected criteria or rules.

2.2.3 Pressure Swing Adsorption Polytope

In chemical industries, operating modes as reversed flow for reaction-regeneration energy efficient coupling of endothermic and exothermic reactions, countercurrent flow and induced pulsing flow in trickle beds, unsteady operations, cyclic processes, extreme conditions, low-frequency vibrations to improve gas–liquid contacting in bubble-columns, high temperature and high pressure technologies, and supercritical media, and use of composite structured packing achieving low pressure drop through vertical stacking of catalyst, are considered for practical application (Ruthven 1984).

Cycling operation methods are of great importance in oil chemistry, in pharmaceutical and food industry, bio-refinery, isotopes separation, hydrogen purification, desalinization, and so forth (Yang 1987). Cyclic separation technologies such as pressure swing adsorption (PSA), temperature swing adsorption (TSA), vacuum swing adsorption, cyclic zone adsorption, simulated moving beds (SMB) chromatography, pressure swing reactor and reverse flow reactor, parameter pumping and so forth, are unsteady non-linear processes difficult to put into practice and to control (Luo 2013).

In pressure swing adsorption (PSA) processes, gas mixtures are separated by selective adsorption over a bed of sorbent materials. The cyclic nature of these processes arises from the high pressure adsorption phase and the subsequent low-pressure regeneration phase. The PSA cycle was accepted for commercial use in air drying.

Thermal swing adsorption (TSA) processes are similar to pressure swing adsorption processes and also separate gas mixture, but here the cyclic nature of these processes arises from the low temperature adsorption phase and the subsequent high temperature regeneration phase. There exist processes that are a combination of PSA and TSA.

A reverse flow reactor is a packed bed reactor in which the flow direction is periodically reversed in order to trap a hot zone within the reactor. In this way even systems with a small adiabatic temperature rise can be operated without preheating the feed stream.

In a pressure swing reactor, reaction and adsorption occur in the same bed.

The adsorption is typically used to purify one of the reaction products. The cyclic nature of a pressure swing reactor arises from the same high pressure adsorption and low pressure regeneration phases as in the pressure swing adsorption.

As the separation complexity and the number of devices increases it becomes very difficult to formulate a feasible schedule much less an optimal one.

The proposed approach allowing operating cyclic separations in multi-component high complexity conditions is that of self-evolvable and smart separation systems. These are systems that can change autonomously the schemes and the dynamic behavior and are capable to control and to take advantage of the unexpected events of their environment in increasingly complex ways. Smart separation systems should have emergent, not entirely pre-programmed, behavior.

Non-stationary and periodically operated separation device with self-evolvability based on schemes modification on self-configuring schemes and multi-scale schemes organized by self-similar replication at different conditioning levels are significant.

Figure 2.23 shows a linear four steps schemes.

Well-known examples are the scheme involving 4-beds and 4-steps (Ruthven 1984; Yang 1987; Luo 2013). The 4 steps are: 1-pressurization 2-saturation 3-depressurization 4-purge in this scheme.

The example of PSA system we may consider is the process of pressure swing adsorption of water from air onto alumina. It is designed to separate the water from the air, so that dry air is obtained.

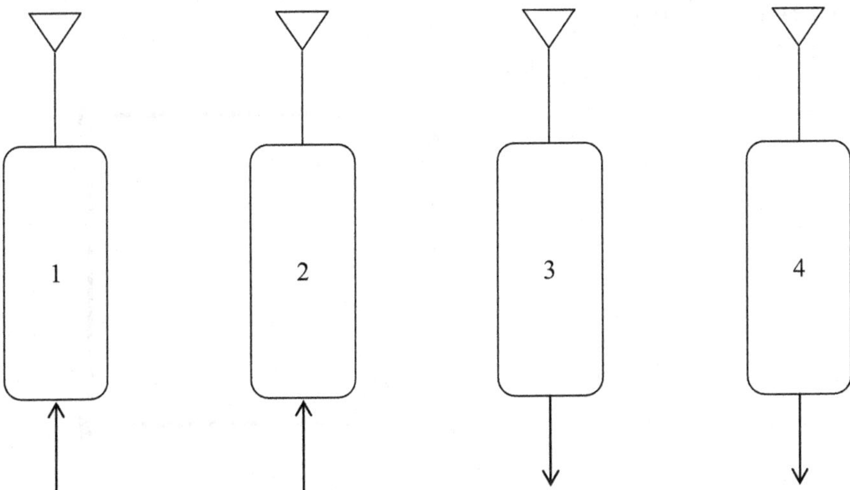

Fig. 2.23 Four steps schemes

The cyclic process is operated in four consecutive steps as shown in Fig. 2.23.

The first step is the adsorption step. In this step the carrier gas, air, with a trace of the adsorbate, water, which is to be removed from the carrier gas, is led into the reactor in which the adsorbent, alumina, adsorbs the adsorbate. At the product end of the reactor the gas stream contains, close to, no adsorbate. During this stage of the process the pressure is maintained at a high level.

Before the adsorbent in the reactor is completely saturated with adsorbate so that it does not adsorb any more adsorbate, the product end of the reactor is closed and the pressure is released at the feed end of the reactor. This is the second step, the blow down step.

When the pressure has dropped to a sufficient low level, it is maintained at this level and clean carrier gas is led into the reactor at the product end so that the adsorbent in the reactor is purged, that is cleaned. This is the third step, the purge step.

When the adsorbent has lost enough of its loading, the product end of the reactor is again closed and the pressure is raised to the old high level. This is the fourth step, the pressurization step. When the pressure has reached its high level the process switches again to the first step.

A sequence of the above described steps is called a cycle. That cycle is a 2D presentation.

The operation steps are denoted as follows:

1-adsorption
2-blow down
3-purge
4-pressurization.

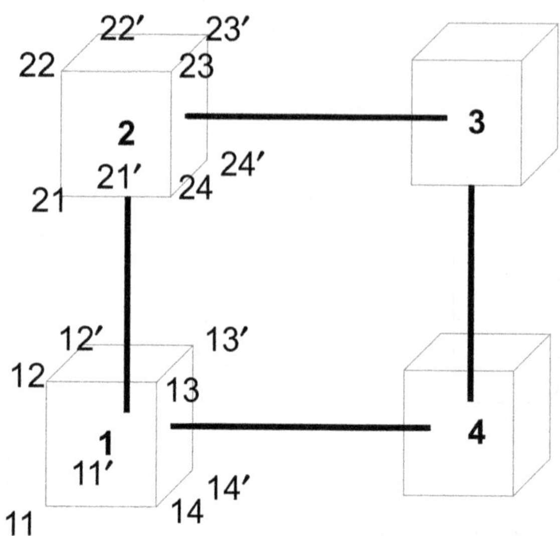

Fig. 2.24 Polytope 5D for PSA separation

For the reverse or modified process the notations are:

1′-modified adsorption
2′-modified blow down
3′-modified purge
4′-modified pressurization.

Figure 2.24 shows a two-scale PSA scheme.

Using the thick hyper-lines, the 5D cube is shown as "square of cubes".

The main steps denoted 1, 2, 3, and 4 may be subdivided by a geometric lower scale having the same four steps. This may be considered as separation schemes of separation schemes.

For instance, 12 mean the step 1 at larger scale but 2 at the smaller scale. The step 12′ reverses or modifies the step 12.

Figure 2.25 shows a 6D polytope for two-scale PSA scheme.

With the thick hyper-lines, the 6D cube is shown as "cube of cubes".

The main steps 1, 2, 3, and 4 are reversed or modified by the main steps 1′, 2′, 3′, and 4′.

2.2.4 *Simulated Moving Bed Polytope*

Figure 2.26 shows a cyclic presentation of SMB. It is a 2D presentation.

The simulated moving bed (SMB) is realized in practice by connecting several single chromatographic columns in series and simulating the movement of solid by

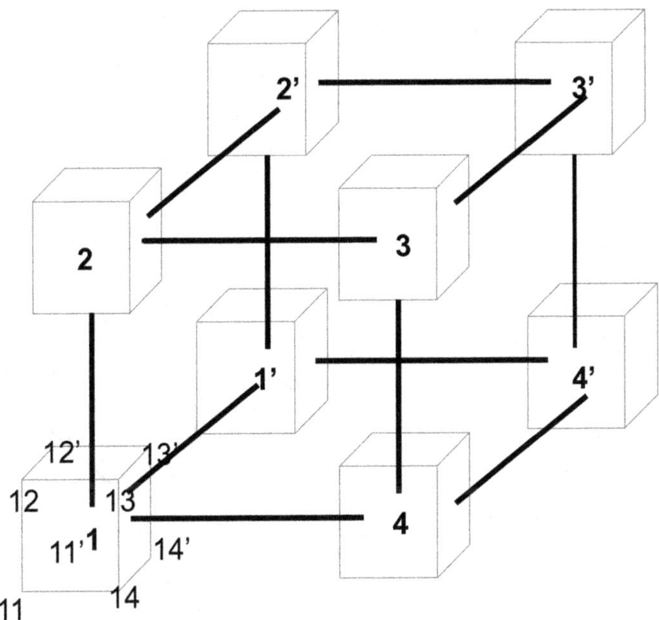

Fig. 2.25 Polytope 6D for PSA separation

cyclic switching of the inlet and outlet ports. The complex dynamics, important non linearity, and multivariable character of the SMB make its control a difficult task.

The SMB consists of four columns or beds interconnected in a circular arrangement.

The positions of the feed, F, extract, E, desorbent, D and raffinate R are changed periodically, in four steps, corresponding to the four columns denoted by 1, 2, 3 and 4.

Notations are:

F-feed

E-extract

D-desorbent

R-raffinate

For reversed or modified steps the notations are:

F′- modified feed

E′-modified extract

D′-modified desorbent

R′-modified raffinate.

Figure 2.27 shows a polytope 5D for SMB.

Using the thick hyper-lines, the 5D cube is shown as "square of cubes".

The main steps F, E, D, and R may be subdivided by a geometric lower scale having the same four steps. It is a cyclic separation scheme of cyclic separation schemes.

Raffinate

1

2

Feed

Port
Switching
Direction

Desorbent

4

3

Extract

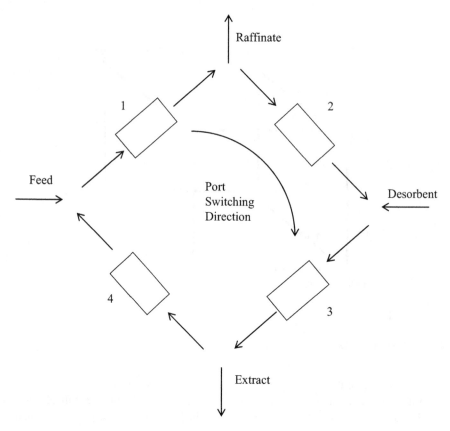

Fig. 2.26 Cyclic presentation for SMB

Fig. 2.27 Polytope 5D for SMB separation

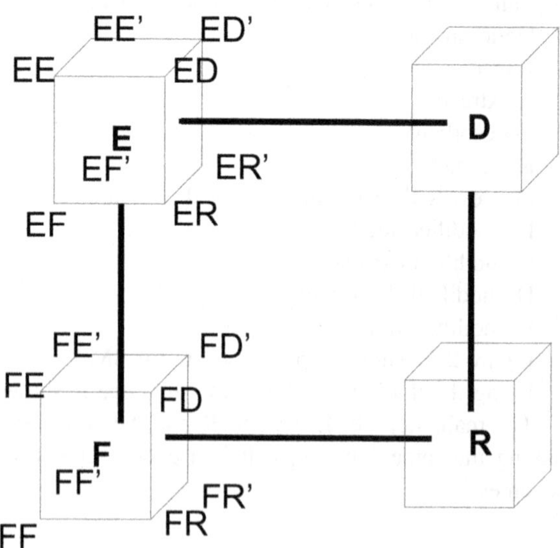

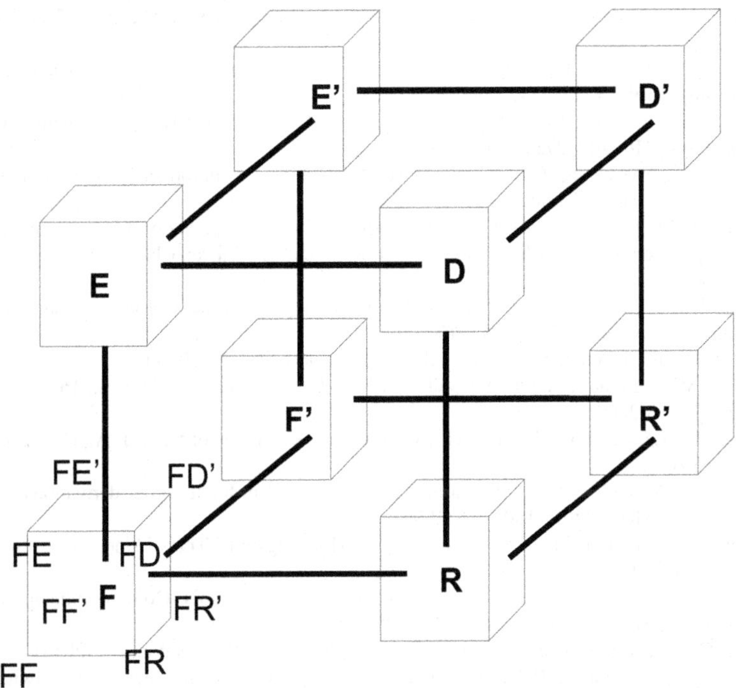

Fig. 2.28 Polytope 6D for SMB separation

For instance, FE means the step F at larger scale but E at the smaller scale.

The step FE′ reverses or modifies the step FE. If we refers to columns 12′ reverses or modifies the step 12.

Figure 2.28 shows a polytope 6D for SMB.

With the thick hyper-lines, the 6D cube is shown as "cube of cubes".

The main steps F, E, D, and R are reversed or modified to the main steps F′, E′, D′, and R′.

The 5D and 6D schemes for PSA or SMB may be interpreted as cycles of cycles.

High-dimensional schemes have been studied as parallel computing architectures (Seitz 1985; Ammon 1998).

There exist a large number of high-dimensional cyclic separations schemes pending to be explored by chemical engineers.

References

Aguiar, M., Sottile, F.: Cocommutative Hopf algebras of permutations and trees. J. Algebraic Comb. **22**(4), 451–470 (2005)

Ammon, J.: Hypercube Connectivity within ccNUMA Architectures. SGI Origin Team, Mountain View, CA (1998)

Brouder, C., Frabetti, A.: QED Hopf algebras on planar binary trees. J. Algebra **267**(1), 298–322 (2003)

Connes, A., Kreimer, D.: Hopf algebras, renormalization and noncommutative geometry. Comm. Math. Phys. **199**(1), 203–242 (1998)

Connes, A., Kreimer, D.: Renormalization in QFT and the Riemann-Hilbert Problem I. Comm. Math. Phys. **210**, 249–273 (2000)

Dăscălescu, S., Năstăsescu, C., Raianu, Ş.: Hopf Algebras. An introduction, Pure and Applied Mathematics, vol. 235 Marcel Dekker (2001)

Demicoli, D., Stichlmair, J.: Novel operational strategy for the separation of ternary mixtures via cyclic operation of a batch distillation column with side withdrawal. Chem. Eng. Trans. **3**, 361–366 (2003)

Flodman, H.R., Timm, D.C.: Batch distillation employing cyclic rectification and stripping operations. ISA Trans. **51**(3), 454–460 (2012)

Fomin, S.: Duality of graded graphs. J. Algebraic Combin. **3**, 357–404 (1994)

Fulman, J.: Mixing time for a random walk on rooted trees. Electron. J. Combin., **16**(1), Research Paper 139, 13(2009)

Grossman, R., Larson, R.G.: Hopf algebraic structures of families of trees. J. Algebra **126**, 184–210 (1989)

Grossman, R., Larson, R.G.: Solving nonlinear equations from higher order derivations in linear stages. Adv. Math. **82**(2), 180–202 (1990)

Hoffman, M.E.: Combinatorics of rooted trees and Hopf algebras. Trans. Amer. Math. Soc. **355**, 3795–3811 (2003)

Hoffman, M.E.: Rooted Trees and Symmetric Functions: Zhao's Homomorphism and the commutative Hexagon. arXiv:0812.2419 (2008)

Iordache, O.: Polytope Projects. Taylor & Francis CRC Press, Boca Raton, FL (2013)

Iordache, O.: Implementing Polytope Projects for Smart Systems. Springer, Cham, Switzerland (2017)

Kotai, B., Lang, P., Balazs, T.: Separation of maximum azeotropes in a middle vessel column. In: Institution of Chemical Engineers Symposium Series, vol. 152, pp. 699–708. Institution of Chemical Engineers, 1999 (2006)

Loday, J.-L., Ronco, M.: Hopf algebra of the planar binary trees. Adv. Math. **139**, 299–309 (1998)

Luo, L. (ed.): Heat and mass transfer intensification and shape optimization: A multi-scale approach. Springer, Berlin (2013)

Pang, C.Y.: Markov Chains from Descent Operators on Combinatorial Hopf Algebras. arXiv preprint arXiv:1609.04312. 2016 Sep 14 (2016)

Qing, Y.: Differential posets and dual graded graphs. Diss. MIT, Cambridge (2008)

Ruthven, D.M.: Principles of Adsorption and Adsorption Processes. Wiley, New York (1984)

Seitz, C.L.: The cosmic cube. Commun. ACM **28**(1), 22–33 (1985)

Sloss, C.A.: Enumeration of walks on generalized differential posets. M.S. Thesis, University of Waterloo, Canada (2005)

Stanley, R.P.: Differential posets. J. Amer. Math. Soc. **1**, 919–961 (1988)

Stanley, R.P.: Enumerative Combinatorics, vol. 1. Cambridge University Press, Cambridge (1997)

Yang, R.T.: Gas Separation by Adsorption Processes. Butherworths, Boston, MA (1987)

Chapter 3
Structuring and Restructuring

3.1 Supramolecular Chemistry

3.1.1 Supramolecular Polytope

Molecular chemistry developed a set of procedures for making or breaking covalent bonds between atoms in an organized mode and has implemented them for constructing complex novel molecules and materials, presenting a range of original properties of interest for science and technology. Beyond molecular chemistry based on the covalent bond, supramolecular chemistry aims at developing highly complex chemical systems from components interacting through noncovalent intermolecular forces. It has grown into a major field of investigation and has rooted numerous developments at its interfaces with biology and physics, leading to the emergence of supramolecular science and technology (Lehn 1995; Ariga and Kunitake 2006).

A supramolecular structure is an organized, complex entity that is created from the association of two or more chemical species held together by intermolecular forces. Supramolecular structures are the result of additive and cooperative interactions, including hydrogen bonding, hydrophobic interactions and coordination.

The supramolecular properties are more than the sum of the properties of each individual component.

Supramolecular chemistry has paved the way for the implementation of the concept of molecular information in chemistry, with the aim of gaining progressive control over the spatial and temporal features of matter and over its complexification through self-organization.

Systems of interests are hydrogen bonding arrays, sequences of donor and acceptor groups, van der Waals shapes, ion coordination sites, and so on. The constructions involved the design and investigation of pre-organized molecular receptors of several types, capable of binding specific substrates with high efficiency and selectivity that is through processes of high information content. Such

O. Iordache, *Advanced Polytopic Projects*, Lecture Notes
in Intelligent Transportation and Infrastructure,
https://doi.org/10.1007/978-3-030-01243-4_3

developments lead to perceiving chemistry as an information science, the science of informed matter, involving an ever clearer perception, deeper analysis, and more deliberate application of the information paradigm in the elaboration and transformation of matter, thus tracing the path from merely condensed matter to more and more self-organized matter towards systems of increasing complexity (Lehn 1999, 2002, 2004; Otto 2003). In chemistry, the language of information is extending that of constitution, structure, and transformation as the field develops towards more and more complex architectures and behaviors.

Molecular recognition relies on design and pre-organization and implements information storage and processing. Investigation of self-organization and self-processes in general, relies on design since it implements programming and programmed systems. The emerging phase, introduces adaptation and evolution, based on self-organization through selection in addition to design, and implements chemical diversity and informed dynamics.

The role of template in the synthesis process opens new opportunities such as spatial control over self-assembly and folding. Examples of self-templating systems are numerous.

We start with a study of disulfide macrocycles formation (Chung et al. 2009; Li et al. 2011).

Figure 3.1 shows a polytope 4D for disulfide macrocycles formation (Iordache 2013).

Figure 3.1 considers the case study of disulfide macrocycles formation and shows the place of different compounds from dynamic combinatorial libraries in the polytope frame (Li et al. 2011).

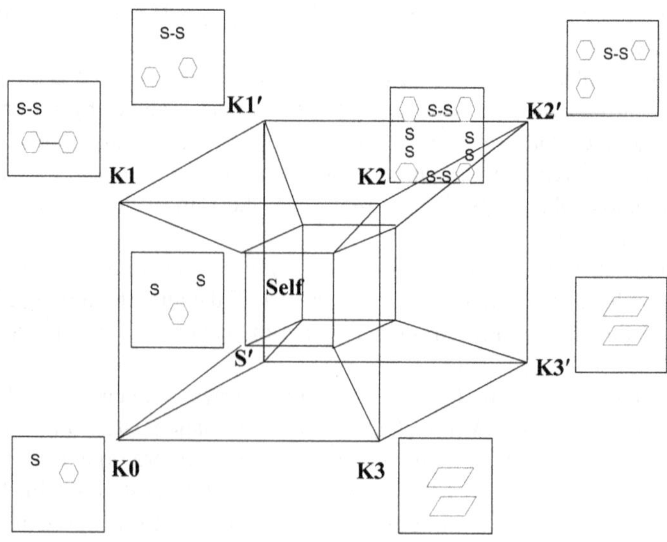

Fig. 3.1 Polytope 4D for disulfide macrocycles formation

The front face of the external cube shown in Fig. 3.1 describes the construction of the supramolecular system.

K0 refers to the substances allowing the synthesis. K1 shows the polymer chains and the dithiol. K2 corresponds to disulfide macrocycles formation. Several possibilities should be considered for different polymerization degrees.

K3 corresponds to assembly of macrocycles. It is a self-templating and self-assembly process. The way K0, K1, K2, K3 describes the supramolecular system construction from small to large.

Such processes appear in the study of folding (Lao et al. 2010; Hunt et al. 2009; Hunt and Otto 2011).

K1 refers to covalent chemistry, while K2 and K3 to different levels of supramolecular chemistry.

The back face of the external cube shown in Fig. 3.1 describes the deconstruction and restructuring of the supramolecular system.

It is the way: K3′→K2′→K′→S′ from large to small. The resulting polytope generalizes the well studied reversible reactions.

Another case study to be considered concerns the construction of helicates.

Initial studies of supramolecular synthesis of helicates are restricted to K0, K1 and K2 levels (Hasenknopf et al. 1996, 1997).

A new level of self-organization can be associated to K3 (Berl et al. 2000; Petitjean et al. 2002).

Figure 3.2 illustrates the polytope 4D for helicates formation.

The front face of the external cube shown in Fig. 3.2 describes the construction of the supramolecular system.

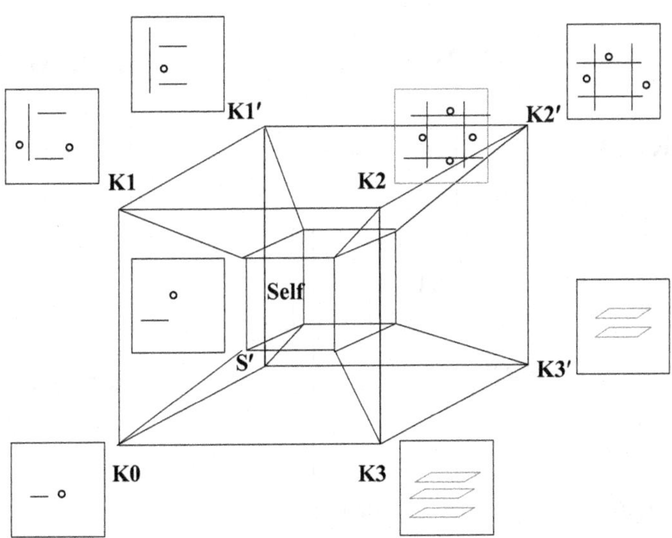

Fig. 3.2 Polytope 4D for helicates formation

The multiplicity ensures the flexibility of the K3 schemes. Layered, stacked structures with weak but multiple linkages between layers allow self-evolvability and smartness. Such structures allow us to create a range of innovative potential realizations due to the weak linkages between layers. We can say that the essential characteristics of self-evolvability and smartness are based on folding.

The thermodynamic templating leads to amplification of just one product and the knowledge gained can aid in the development of host-guest systems, chemical receptors and new ligand systems. Significant routes for the amplification of one particular product include foldamers, where the three dimensional configuration of one product held together by non-covalent interactions is much more stable than the others and selection self-assembly, where the components of the library aggregate to form large molecules through non-covalent interactions, that is supramolecular assemblies, where one assembly is more stable than the other possible candidates.

The "Self" should coordinate and facilitate the two ways in duality relation.

Figure 3.3 shows a general polytope 4D for supramolecules formation.

The notations are:

- C-components
- M-molecules
- SM-supramolecules
- SSM-supra supramolecules.

The back face of the external cube of the 4D polytope shows the modified or restructured compounds, at different levels of association.

The construction or structuring process starts from low number of separated compounds and self-evolves towards the supra supramolecules. It is the way: S→K1→K2→K3 from small to large.

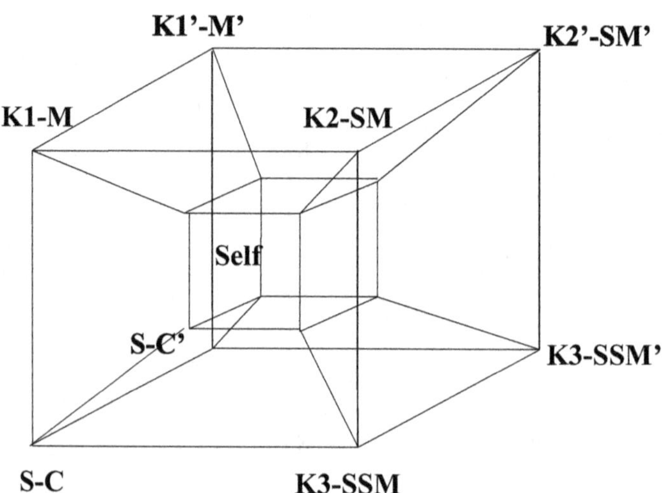

Fig. 3.3 Polytope 4D for supramolecules formation

The deconstruction or restructuring describes evolution from supra supramolecules towards separated compounds.

It is the way: K3′→K2′→K′→S′ from large to small.

3.1.2 Higher Dimensions

Observe that any level K0, K1, K2 or K3 may be studied in more detail as polytopic architecture of sublevels.

This corresponds to the cases when at different levels there are structured network of interacting polymers as for instance the disulfide macrocycles or helicates.

Chemical systems study shows that polymers can self-assemble more or less randomly, and by chance form autocatalytic cycles of supramolecular systems. Inspired by hypercycle theory a cyclic system of supramolecular cycles is proposed here. The supramolecular cycles are functionally related self-replicative supramolecular units formed into multiple feedback loops.

Figure 3.4 shows the polytope 5D for a system of supramolecular systems.

Only four supramolecular unit systems SM0, SM1, SM2, and SM3 have been considered here but this number may be different.

SM0, SM1, SM2 and SM3 denote supramolecular unit systems, organized as a cycle.

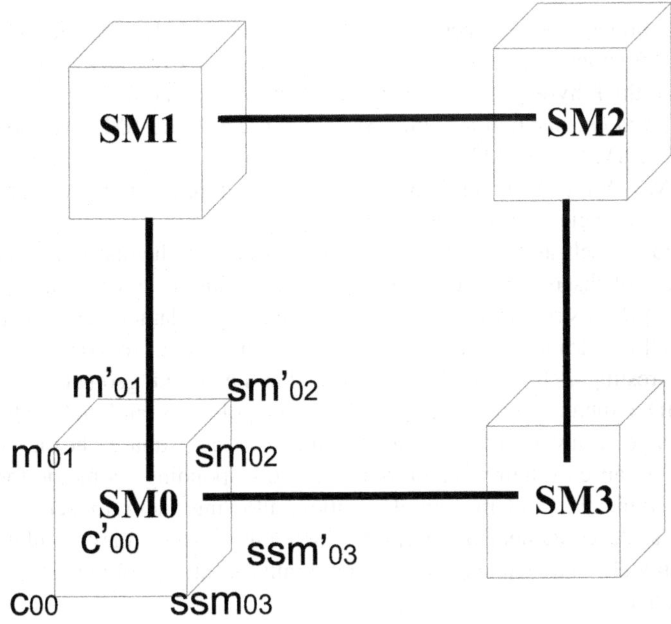

Fig. 3.4 Polytope 5D for supramolecular cycles system

They interact as shown in Fig. 3.4. SM0 contributes to SM1 structuring SM1 contributes to SM2 structuring and so on. SM3 close the loop and contributes to SM0 structuring in a kind of autocatalytic behavior. The cyclic organization of the system of supramolecular unit systems ensures the general structure stability.

The thick hyper-lines describe the complex interactions between levels and molecules.

Every supramolecular system in the cycle is represented as a cube. The entire system is represented as a square of cubes. It may be considered as a supramolecule of supramolecules.

The idea is that of iterative constructions of systems of systems over systems and so on. The systems of different levels of reality show specific properties.

- SMj, j = 0, …, 3 are the supramolecular cycle elements.
- cj0-components for SMj at the level 0
- mj1-molecules for SMj at the level 1
- smj2-supramolecules for SMj at the level 2
- ssmj3-supra supramolecules for SMj at the level 3.

The back face of the cubes outlines the restructured or modified compounds. The notations are

- cj0'-restructured components for SMj at the level 0
- mj1'-restructured molecules for SMj at the level 1
- smj2'-restructured supramolecules for SMj at the level 2
- ssmj3'-restructured supra supramolecules for SMj at the level 3.

The component supramolecular systems are: SM0, SM1, SM2 and SM3.

Figure 3.5 shows the polytope 6D for supramolecular cycles systems.

With the thick hyper-lines, the 6D cube is shown as "cube of cubes".

The supramolecular unit systems: SM0, SM1, SM2 and SM3 are modified to: SM0′, SM1′, SM2′ and SM3′.

Here SM0′, SM1′, SM2′ and SM3′ denote restructured supramolecular systems, organized as a cycle in a parallel plane.

The study of self-assembling systems in coordination chemistry has delivered a wide variety of discrete molecular architectures, including twisted structures (e.g. helicates), latticed structures (e.g. grids, racks, ladders), filamentous structures (e.g. rods, metallodendrimers), and interlaced structures (e.g. rotaxanes, catenanes, knots). Interesting architectures are to be found in the closed, geometrically-shaped, 2D and 3D structures known as metallocyclic polygons and polyhedra. These materials have been variously termed molecular triangles, squares, hexagons, cubes, and so on, because of their resemblance to the corresponding geometric form. Such molecules typically contain central cavities, allowing them to act as artificial molecular scale containers or receptors. In doing so, they may exhibit unusual electrochemical, magnetochemical, photoluminescent, catalytic, or synthetic chemical effects.

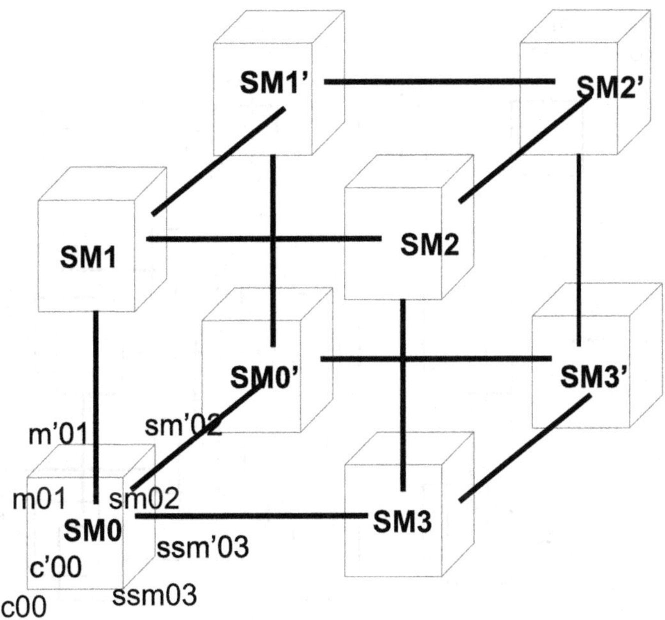

Fig. 3.5 Polytope 6D for supramolecular cycles system

Figure 3.6 illustrates the formation of polytopes 6D for organo metallic supramolecular compounds.

An example is the reaction over several weeks of [([9]ane-S_3)Ru(DMSO)Cl_2] with 4,4′- bipyridine resulting in the formation of the cube which linked by twelve of the bipyridine ligands. DMSO denotes dimethyl sulfoxide (Swiegers and Malefetse 2002).

Structures other than a cube have been inconsistent with the crystallographic, mass spectral and NMR data. Three oxidation waves in cyclic voltammetry, rather than the one which may have been anticipated suggest that an interaction between the metal ions therefore exists. Interactions of this type raise the possibility of novel molecular devices, such as multiple-state switches or tunable sensors. The host–guest chemistry of such boxes may influence electron and energy transfer processes. The eight lines connecting a pair of two 3D cubes are represented in Fig. 3.6 by thick hyper-lines. These describe the interactions of the eight Ru^{II} corner ions.

With the thick hyper-lines, the 6D cube is shown as "cube of cubes".

The diversity of resulting supramolecular systems is explained by the interaction of assembly and disassembly processes. Since complementary processes may be essentially irreducible the integration or coordination of complementary systems requires a higher level system, a new dimension that emerges as a new hierarchical level of complexity and a new dimension. The more complex a system is, the more dimensions are necessary for comprehension and description.

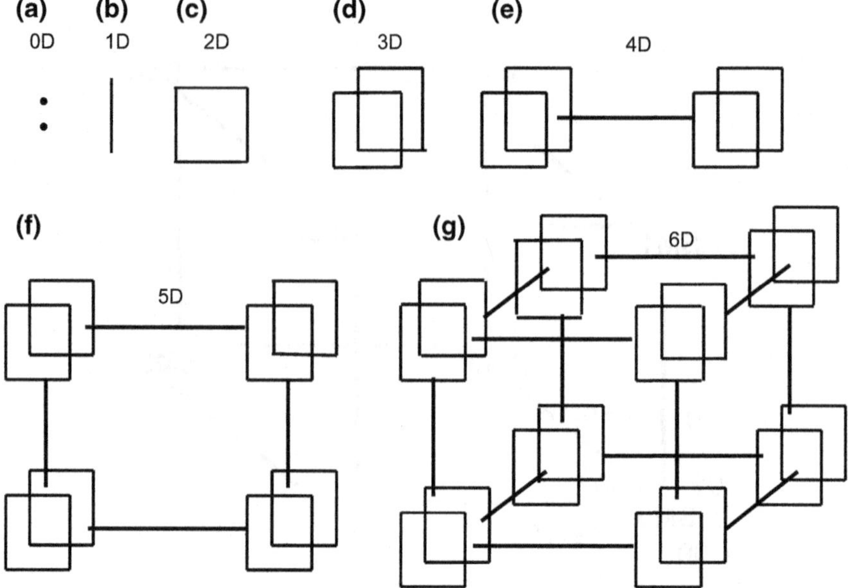

Fig. 3.6 Polytopes 6D for organo metallic compound formation

The swing between assembly and disassembly steps imposes that supplementary dimension. Correlating the constructing and deconstructing processes helps us make supramolecules that could not be made by only one of these processes considered separately.

It is important to have in mind what kind of architectures to use when forming new structures and organizations out of given collections of molecules.

3.2 Quasi-species

3.2.1 Hypercycles

A notable approach in the study of genetic code evolution is the Eigen's model of hypercycles systems of mutually autocatalytic components. It consider the question of under what conditions, the system can self-organize to a dynamic stability (Eigen 1971; Eigen and Schuster 1979). The approach was based on the view that the self-organization including the development of hypercycles is a process that can occur in a homogeneous system by intrinsic necessity.

Eigen and Schuster proposed the model of hypercycles as a hypothetical stage of macromolecular evolution.

The hypercycle is a self-reproducing macromolecular system, in which RNAs and enzymes cooperate in the following manner: there are RNA matrices (I_i); i-th RNA codes i-th enzyme E_i ($i = 0, 1, 2, \ldots, n$); the enzymes cyclically increase RNA's replication rates, namely, E_1 increases replication rate of I_2, E_2 increases replication rate of I_3, …, E_n increases replication rate of I_0. This completes the cycle. In addition, the mentioned macromolecules cooperate to provide primitive translation abilities, so the information, coded in RNA-sequences, is translated into enzymes, analogously to the usual translation processes in biological objects. The cyclic organization of the hypercycle ensures its structure stability. For effective competition, the different hypercycles should be placed in separate compartments.

The replication enzymes ensure the more accurate RNAs' replication as compared with quasispecies, providing opportunities for further macromolecular structure improvements. Eigen and Schuster consider hypercycles as predecessors of protocells, the primitive unicellular biological organisms.

Developing the hypercycle model, Eigen and Schuster discussed the difficult problem of how could the real very complex translation mechanism and unique genetic code be created during macromolecular self-organizing process. Plausible evolution steps were outlined and a corresponding well-defined mathematical model was developed.

Eigen and Schuster considered that the primitive genetic code may use units of less than three bases. During its early evolution, the code would have increased both the number of codons and the coded amino acids and the present code would reflect the pattern of this historical expansion.

In the view of Kuhn and Waser (1994), understanding the origin of living systems is a physical problem: to find a sequence of physicochemical stages, beginning with prebiotically reasonable conditions and leading to self-organization of matter and to systems equipped with a like-alive genetic apparatus (Kuhn and Waser 1994; Kuhn and Kuhn 2003) Life and evolution are considered as physical phenomena, belonging to physics and molecular engineering.

The genome generates different dynamical systems that promotes their stability and survive and in that way serves as seeds of a generally self-evolvable system

Figure 3.7 illustrates the cyclic schemes associated to hypercycles.

The hypercycle is a self reproducing macromolecular system in which RNAs and enzymes cooperate. There are RNA matrices (I_i), the i-th RNA codes the i-th enzyme E_i. The enzymes cyclically increase RNA's replication rates, namely,

The mentioned macromolecules cooperate to provide primitive translation abilities, so the information, coded in RNA-sequences, is translated into enzymes analogous to the usual translation processes in biosystems.

The cyclic organization of the hypercycle ensures its structure stability. For effective competition, the different hypercycles should be placed in separate compartments.

Figure 3.7 shows 5D schemes for hypercycles.

I0, I1, I2, and I3 denotes the RNA matrices systems, E0, E1, E2, and E3 the enzyme systems.

The basic RNA matrices are: I0, I1, I2, and I3.

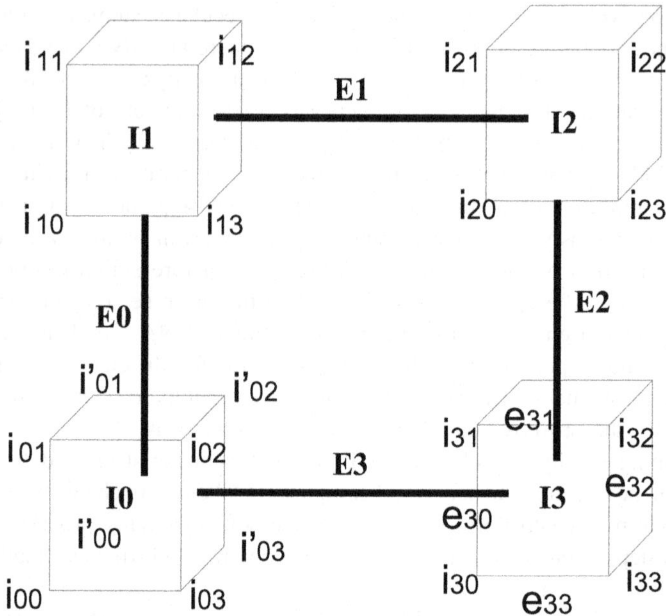

Fig. 3.7 Schemes 5D for hypercycles

Using the thick hyper-lines, the 5D cube is shown as "square of cubes".

Figure 3.7 shows that some RNA may induce the reproduction of other metabolites in cyclic manner. Supposing that I3 is in this situation, e_{30} increases replication rate of i_{31}, e_{31} increases replication rate of i_{32}, e_{32} increases replication rate of i_{33}, and e_{33} increases replication rate of i_{30}. The number of RNAs in each cycle may vary.

Metabolism is reproduction of invertible macromolecules (proteins, RNA, and DNA), which in turn catalyze the reproduction of other metabolites in a cyclic manner. Each cube in Fig. 3.7 represents a cyclic metabolic pathway reproducing a metabolite of another cyclic metabolic pathway, and so on. The number of nodes in each cycle may vary. This is a general representation of autocatalytic hypercycles. It may be considered as a hypercycle of hypercycles.

3.2.2 Quasi-species Polytope

3.2.2.1 Sequence Space

One way to study the diverse nucleotide sequences in the genes of viruses is to map them into a multidimensional matrix called a sequence space (Eigen 1993, 2000).

In this space, each point represents a unique sequence, and the degree of separation between points reflects their degree of dissimilarity. The space can be most easily drawn for short sequences consisting of binary digits. For a sequence with just one position, there are only two possible sequences, and they can be drawn as the end points of a line. For a sequence with two positions, there are four permutations, which form the corners of a square. The variations on a three-digit sequence become the corners of a cube, and the variations on a four-digit sequence are the vertices of a 4D hypercube.

Figure 3.8 illustrates steps for sequence construction. Each high-dimensional space is built iteratively by drawing the previous diagram twice and connecting the corresponding points.

The sequence spaces for viral genomes are far more complex than these simple figures because they involve thousands of positions that can each be occupied by one of four different nucleotides.

The construction of a high-dimensional sequence space was illustrated by Eigen‘s studies (Eigen et al. 1988). Each additional sequence position adds another dimension, doubling the diagram for the shorter sequence.

Figure 3.8 shows elements of the sequence space from 0D to 5D.

With the thick hyper-lines, the 5D cube is shown as “square of cubes”.

Figure 3.9 shows the sequence space 6D.

With the thick hyper-lines interpretation, the 6D cube is shown as “cube of cubes”.

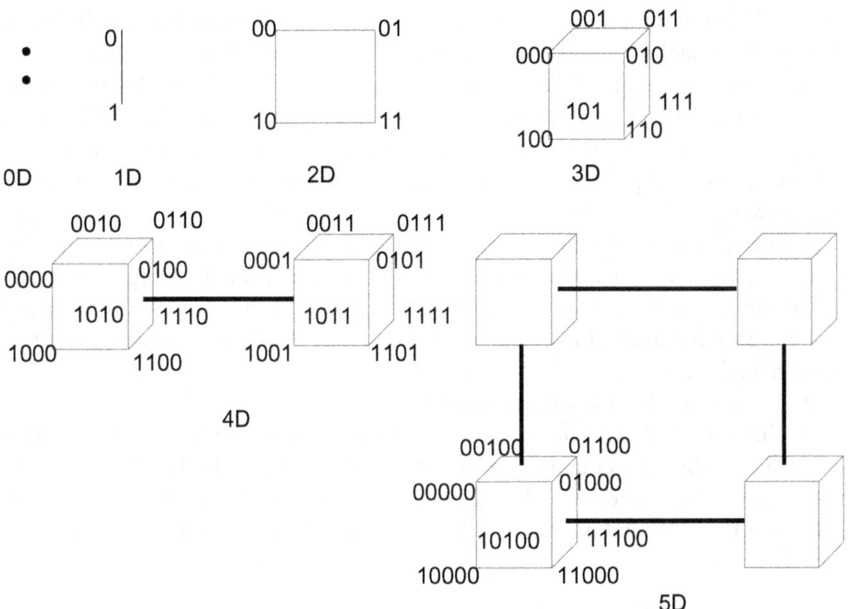

Fig. 3.8 Sequence space 5D

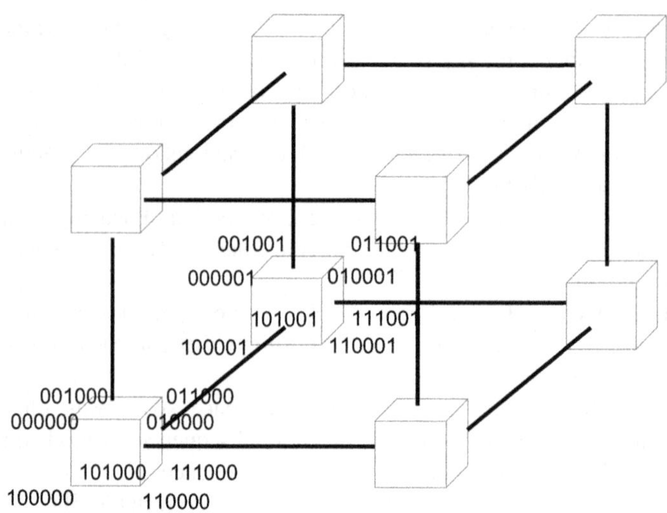

Fig. 3.9 Sequence space 6D

3.2.3 Doubling and Contracting

A generalization of the sequence space construction proposed by Eigen (1993) is to consider the doubling and the contraction of intervals of lattices diagrams.

Doubling is a way to produce new structures from older ones. This means that we can compose and produce new structures on a given level, by replicating at lower levels. The rules may vary and be flexible due to the relevant context.

Constructive properties of lattice are of interest for the study of lattice boundedness.

A lattice is bounded if it can be constructed, starting with the one-element lattice, by applying a finite sequence of a simple operation called interval doubling (Day 1970).

This operation assigns to a poset P and an interval I, a new poset $P' = P[I]$ by doubling in P the interval I, that is by replacing I in P with the direct product by a two-element lattice.

Denote by "+" the disjoint set union.

The interval doubling construction is defined as follows (Caspard et al. 2004):

Let $(P, \leq)$ be a poset and $I \subseteq P$ an interval of P. Denote by $B = (\{01\} \leq)$ the two-element lattice where $0 < 1$. The poset P' defined on the set $(P - I) + (I \times B)$ is denoted $P' = P[I]$ and is given by the following relation of order: $x' \leq y'$ if and only if:

- $x',\in P - I$ and $x' \leq y'$ or
- $x' \in P - I$, $y' = yi \in I \times B$ and $x' \leq y'$ or
- $x' = xi \in I \times B$, $y' \in P - I$ and $x' \leq y'$ or
- $x' = xi \in I \times B$, $y' = yj \in I \times B$, $x \leq y$ and $i \leq j$ in B.

A lattice L is bounded if there exists a sequence $B = L_1, \ldots, L_i, \ldots, L_p = L$ of lattices and a sequence $I_1, \ldots, I_i, \ldots, I_{p-1}$ such that Ii is an interval of L_i and $L_{i+1}=L_i[I_i]$ for every $i < p$.

Figure 3.10 shows a series of interval doublings starting with the two-element lattice.

Figure 3.11 shows a single state doublings. The state 010 is doubled to the segment 0101−0100.

The lattices L1, L2, L3 and L4 are bounded.

Since a bounded lattice is a lattice which can be constructed starting from B by a finite sequence of interval doublings, such a lattice is equally characterized by the fact that it can be contracted until B by an iteration of the operation opposite to the interval doubling. We can call this operation, interval contraction.

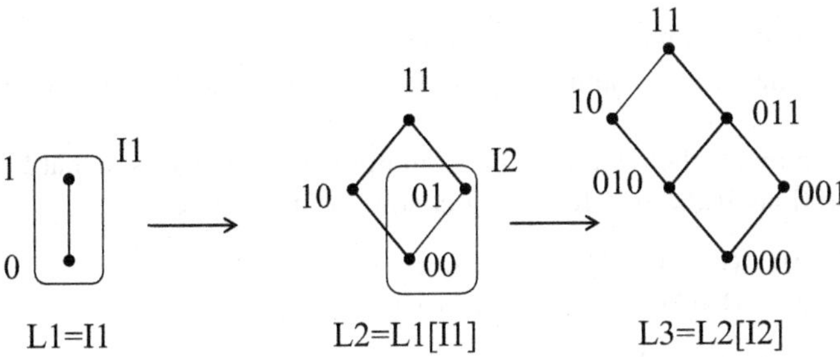

Fig. 3.10 Interval doubling

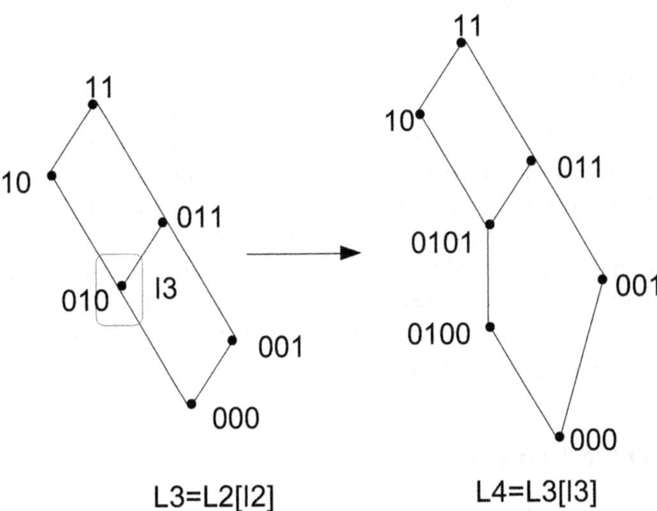

Fig. 3.11 Single state doubling

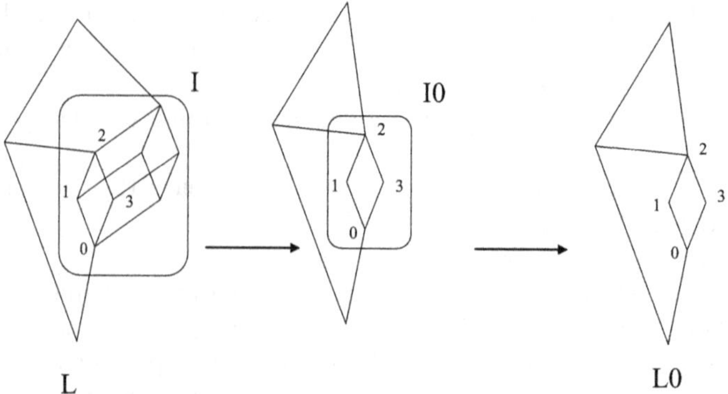

Fig. 3.12 Interval contracting

First we need to define the gluing conditions:

Let I be an interval of a lattice L, with I equal to the direct product of an interval I0 by B.

We denote by I1, the interval I- I0, isomorphic with I0. We say that I satisfy the gluing conditions if the two following conditions are verified:

- $\forall$(y, x1, x0) $\in$(L-I1) $\times$ I_1 $\times$ I_0. (y < x1 $\rightarrow$ y $\leq$ x0)
- $\forall$(z, x1, x0) $\in$(L-I0) $\times$ I1 $\times$ I0. (z > x0 $\rightarrow$ z $\geq$ x1).

Let L be a lattice and I $\subseteq$ L an interval of L. We say that I is contractible in L if the two following conditions hold:

- I is equal to the direct product of an interval I0 by B. I1 denotes the interval composed of the elements of I-I0.
- The gluing conditions are satisfied on I.

Figure 3.12 illustrates the interval contracting.

Let L be a lattice and I $\subseteq$ L a contractible interval of L. We call contraction of I in L the operation of constructing a smaller lattice L0 by replacing I with I0 in L.

The contraction of an interval is the inverse operation to the interval doubling.

Figure 3.12 shows the contraction of the interval I of the lattice L to the interval I0.

For the lattice L the interval I is contractible to I0.

Doubling and contracting allows obtaining a large variety of lattices generalizing the action of operators U and D defined for differential posets (Stanley 1988, 1990).

Doubling is associated to operator U, to construction while contracting corresponds to operator D or deconstruction.

3.3 Oriented Syntheses

Diversity-oriented synthesis, DOS, involves the deliberate, simultaneous and efficient synthesis of more than one target compound in a diversity-driven approach to answer a complex problem (Spring 2003; Burke and Schreiber 2004; Anoh et al. 2015).

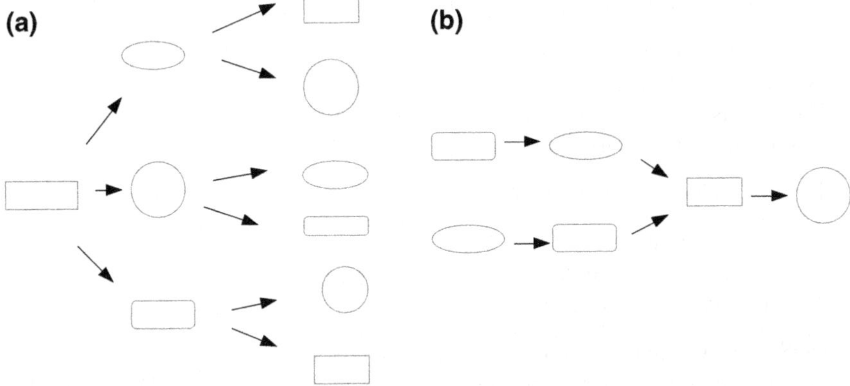

Fig. 3.13 Oriented syntheses

Complexity refers to binding, catalysis, phenotypic and synergistic effects. DOS describes a process whereby diverse collections of complex small molecules are synthesized. In designing a DOS, analysis is performed in a forward sense and a strategy is developed whereby simple starting materials can be transformed into diverse and complex products. Unlike, in target oriented synthesis, TOS, where retro-synthetic analysis allows a complex product to be deconstructed in a backward sense. There is also a difference in the outcomes and goals of both approaches. TOS aims to synthesize a molecule at a discrete point in chemical space. DOS aims to cover as diffuse an area as possible. We would therefore like to suggest that DOS is a more evolved version of combinatorial chemistry. Thus, these terms are not mutually exclusive and the technologies overlap. DOS, however, does differ from traditional combinatorial chemistry as DOS does not target as selected an area of chemical space. An example of this selected targeting in traditional combinatorial chemistry would be in lead optimization for drug discovery. This comparison also serves to highlight another important issue, the subjectivity of diversity.

When a compound collection is synthesized, since the composite molecules are not identical, diversity, to a greater or lesser extent, is incorporated; the racemic synthesis of enantiomers could even be classified as a DOS. Considering diversity as a spectrum may be useful. In one extreme of the molecular diversity spectrum would be where maximal chemical space coverage has been achieved and, in the other extreme, would be a TOS.

Figure 3.13 illustrates the oriented syntheses.

Figure 3.13a illustrates the DOS while Fig. 3.13 illustrates the TOS.

It should be emphasized that DOS and TOS are different strategies with different goals. The above serves to compare the diversity achieved using either approach regardless of the eventual aim; it is not to be implied that DOS is better than TOS as it generates more diversity, merely that, to maximize chemical space coverage, skeletal diversity is essential. It is this skeletal diversity that can be incorporated using the DOS, and not the traditional combinatorial chemistry, approach to library synthesis.

The main characteristics of DOS are:

- Divergent strategy
- No target is specified
- The aim is to maximize diversity and complexity
- Based on forward synthetic analysis.

The main characteristics of TOS are:

- Convergent strategy
- A specific molecules is the target of the synthesis
- Based on retro-synthetic analysis.

DOS and TOS may be considered as complementary.

The integration or coordination of complementary descriptions DOS and TOS requires a higher level description, a new dimension that emerges as a new hierarchical level of complexity, as a center and a higher dimension. The more complex is a system the more dimensions emerge and the polytopic architecture is necessary to manage complexity.

The change of orientation between constructing DOS and deconstructing TOS steps implies that supplementary dimension. Correlating the constructing and deconstructing processes allows us make molecules that could not be made by only one of these processes. Correlation of DOS and TOS is the proposed strategy and drug discovery and personalized drug delivery is a potential domain of application.

In the coming decades a confluence of wireless networks and lab-on-chip sensor technology with application in health monitoring is expected. In such lab-on chip network each sensor node is endowed with a limited supply of chemicals. The network will collectively or via the self-evolution level decide how the drug resources will be spent. Environmental monitoring and improving, new drugs may be obtained by autonomous experimentation architectures correlating DOS and TOS steps (Symes et al. 2012).

References

Anoh, V., Agbo, S., Swande, P.: Exploring the benefit of diversity oriented synthesis (DOS) vis-à-vis other synthetic tools–a review. Chem. Sci. Rev. Lett. **4**(16), 1148–1152 (2015)

Ariga, K., Kunitake., T.: Supramolecular Chemistry—Fundamentals and Applications. Springer, Heidelberg (2006)

Berl, V., Krische, M., Huc, I., Schmutz, M., Lehn, J.-M.: Template-Induced and molecular recognition-directed hierarchical generation of supramolecular assemblies from molecular strands. Chem. Eur. J. **6**(11), 1938–1946 (2000)

Burke, M.D., Schreiber, S.L.: A planning strategy for diversity-oriented synthesis. Angew. Chem. Int. Ed. **43**(1), 46–58 (2004)

Caspard, N., Conte, Le, de Poly-Barbut, C., Morvan, M.: Cayley lattices of finite Coxeter groups are bounded. Adv. in Appl. Math. **33**(1), 71–94 (2004)

Chung, M.-K., White, P.S., Lee, S.J., Gagné, M.R.: Synthesis of Interlocked 56-membered rings by dynamic self-templating. Angew. Chem. Int. Ed. **2009**(48), 8683–8686 (2009)

Day, A.: A simple solution to the word problem for lattices. Canad. Math. Bull. **13**, 253–254 (1970)

Eigen, M.: Selforganization of matter and the evolution of biological macromolecules. Naturwissenschaften **58**(10), 465–523 (1971)

Eigen, M.: Viral quasispecies. Sci. Am. **269**, 42–49 (1993)

Eigen, M.: Viruses: evolution, propagation and defense. Nutr. Rev. **58**(2), 5–16 (2000)

Eigen, M., Schuster, P.: The Hypercycle a Principle of Natural Self-organization. Springer, Berlin (1979)

Eigen, M., McCaskill, J., Schuster, P.: Molecular quasispecies. J. Phys. Chem. **92**, 6881–6891 (1988)

Hasenknopf, B., Lehn, J.-M., Kneisel, B.O., Baum, G., Fenske, D.: Self-assembly of a circular double helicate. Angew. Chem. Int. Ed. Engl. **35**(1), 838–1840 (1996)

Hasenknopf, B., Lehn, J.M., Boumediene, N., Dupont-Gervais, A., Van Dorsselaer, A., Kneisel, B., Fenske, D.: Self-assembly of tetra- and hexanuclear circular helicates. Am. Chem. Soc. **119**, 10956–10962 (1997)

Hunt, R.A.R., Ludlow, R.F., Otto, S.: Estimating equilibrium constants for aggregation from the product distribution of a dynamic combinatorial library. Org. Lett. **11**(22), 5110–5113 (2009)

Hunt, R.A.R., Otto, S.: Dynamic combinatorial libraries: new opportunities in systems chemistry. Chem. Commun. **47**, 847–858 (2011)

Iordache, O.: Polytope Projects. Taylor & Francis CRC Press, Boca Raton, FL (2013)

Kuhn, H., Kuhn, C.: Diversified world: drive of life's origin? Angew. Chem. Int. **42**, 262–266 (2003)

Kuhn, H., Waser, J.: Hypothesis on the origin of genetic code. FEBS Lett. **352**, 259–264 (1994)

Lao, L.L., Schmitt, J.-L., Lehn, J.-M.: Evolution of a constitutional dynamic library driven by self-organization of a helically folded molecular strand. Chem. Eur. J. **16**, 4903–4910 (2010)

Lehn, J.-M.: Supramolecular Chemistry: concepts and perspectives. Wiley-VCH, Weinheim (1995)

Lehn, J.-M.: Dynamic combinatorial and virtual combinatorial libraries. Eur. J. Chem. **5**(9), 2455–2463 (1999)

Lehn, J.-M.: Toward complex matter: supramolecular chemistry and self-organization. Proc. Natl. Acad. Sci. U.S.A. **99**, 4763–4768 (2002)

Lehn, J.-M.: Supramolecular chemistry: from molecular information towards self-organization and complex matter. Rep. Prog. Phys. **67**, 249–265 (2004)

Li, J., Carnall, J.M.A., Stuart, M.C.A., Otto, S.: Hydrogel formation upon photoinduced covalent capture of macrocycle stacks from dynamic combinatorial libraries. Angew. Chem. Int. Ed. **50** (36), 8384–8386 (2011)

Otto, S.: Dynamic combinatorial chemistry: a new method for the selection and preparation of synthetic receptors. Curr. Opin. Drug Discov. Dev. **6**, 509–520 (2003)

Petitjean, A., Cuccia, L.A., Lehn, J.-M., Nierengartenm, H., Schmutz, M.: Cation-promoted hierarchical formation of supramolecular assemblies of self-organized helical molecular components. Angew. Chem. **41**(7), 1195–1198 (2002)

Spring, R.: Diversity-oriented synthesis; a challenge for synthetic chemists. Org. Biomol. Chem. **1** (22), 3867–3870 (2003)

Stanley, R.P.: Differential posets. J. Amer. Math. Soc. **1**, 919–961 (1988)

Stanley, R.P.: Variations on differential posets. In: Invariant Theory and Tableaux. IMA Vol. Math. Appl. 19, pp 145–165. Springer, New York (1990)

Swiegers, G.F., Malefetse, T.J.: Classification of coordination polygons and polyhedra according to their mode of self-assembly. 2, Review of the literature. Coord. Chem. Rev. **225**(1–2), 91–121 (2002)

Symes, M.D., Kitson, P.J., Yan, J., Richmond, C.J., Cooper, G.J., Bowman, R.W., Vilbrandt, T., Cronin, L.: Integrated 3D-printed reactionware for chemical synthesis and analysis. Nature Chemistry. **4**(5), 349–354 (2012)

Chapter 4
Forward and Backward

4.1 Mixing and Growths

4.1.1 Mixing Processes

Differential and integral models have been among the most fundamental tools for the study of mixing processes.

An example is the diffusion equation for the probability density p(x, t).

$$\frac{\partial p}{\partial t} = -\frac{\partial vp}{\partial x} + \frac{\partial^2 Dp}{\partial x^2} \tag{4.1}$$

Here p(x, t) denotes the probability for a particle to be in the position x at the moment t, v denotes the velocity and D the diffusion coefficient.

This equation is called the forward one. We also encounter the backward diffusion equation:

$$\frac{\partial p}{\partial t} = v\frac{\partial p}{\partial x} + D\frac{\partial^2 p}{\partial x^2} \tag{4.2}$$

This is considered equivalent to the forward one, and is useful to compute first passage times and absorption to boundaries probabilities.

These two equations, due to Kolmogorov, have a profound asymmetry, in the forward one, the velocity v(x) and diffusion coefficient D(x) are part of the derivations, whether in the backward one, they are out of it.

The correlation of forward and backward diffusion models for imposed boundaries and initial conditions represents a challenge.

The forward Kolmogorov eq. is useful to describe the residence time distribution, RTD.

Complementing this is the backward Kolmogorov eq. useful in the study of the traveling time distribution TTD.

O. Iordache, *Advanced Polytopic Projects*, Lecture Notes in Intelligent Transportation and Infrastructure,
https://doi.org/10.1007/978-3-030-01243-4_4

An open problem was the correlation of these two complementary models (Iordache 1974, 1976).

The swing between forward and backward steps implies a supplementary dimension for mixing. It is expected that the correlation of the forward and backward processes will allow mixing operations that could not be performed by only one of these processes considered separately.

A simplified model of mixing processes take into account only the particle velocity, more exactly two processes, forward with velocity +v and backward with velocity −v and their interaction (Kac 1974).

Let us denote $p_1(x, t)$ and $p_2(x, t)$ the densities of probabilities for the processes forward (1) and backward (2) and by q the rate of change from a process to another. The two processes are described by the Eqs. (4.3) and (4.4)

$$\frac{\partial p_1}{\partial t} = -v\frac{\partial p_1}{\partial x} - qp_1 + qp_2 \tag{4.3}$$

$$\frac{\partial p_2}{\partial t} = v\frac{\partial p_2}{\partial x} - qp_2 + qp_1 \tag{4.4}$$

Equation (4.3) shows that the probability of the process (1) varies in time due to the convection of the particles of type (1), due to the fact that some particles left the type (1) and other particles of type (2) shift to the type (1).

Equation (4.4) shows that the probability of the process (2) varies in time due to the convection of the particles of type (2), due to the fact that some particles left the type (2) and other particles of type (1) shift to the type (2).

Numerous studies compare the solutions of the parabolic models (4.1, 4.2) to that of a hyperbolic model (4.3, 4.4). The hyperbolic model approximates the dispersion parabolic one with the dispersion coefficient:

$$D = v^2/2q \tag{4.5}$$

The hyperbolic model and its numerous generalizations have become the subject of intense researches provided both by theoretical importance and fruitful applications in statistical physics, transport phenomena in physical and biological systems, hydrology, financial modeling and some other fields.

Some combinatorial results applicable to mixing studies will be presented in the following.

Figure 4.1 shows a sample path for a particle describing the simplified mixing process. Here x denotes the space travelled by the particle. Forward jumps increase x while backward jumps decrease x. Let us suppose that the velocities are v = 1 or v = −1.

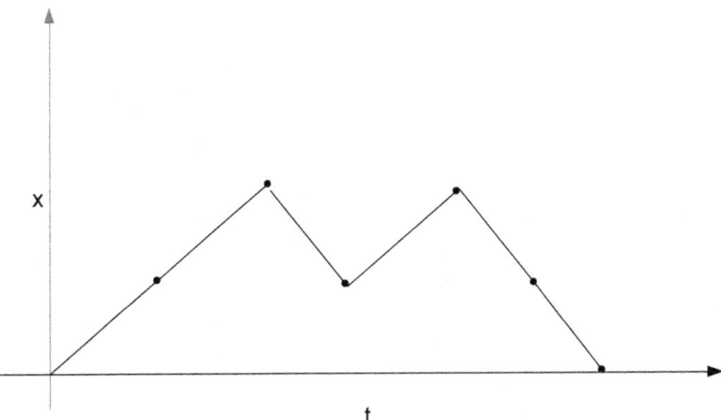

Fig. 4.1 Sample path

We will consider the paths between two visits of the initial position, x = 0.

In the study of mixing the distribution of these visits is as significant as the residence time distribution for chemical reactors design.

These paths may be organized in the frame of Tamari lattices (Geyer 1994; Hivert et al. 2005).

Tamari lattice, was presented as a partially ordered set in which the elements consist of different ways of grouping a sequence of objects into pairs using parentheses (Iordache 2017).

In the Tamari lattice, one group is considered before another if the second group may be obtained from the first by only rightward applications of the associative law:

$$(a\,b)c = a(b\ \ c)$$

In this partial order, any two groupings g_1 and g_2 have a greatest common predecessor, the meet $g_1 \wedge g_2$, and a least common successor, the join $g_1 \vee g_2$.

The Hasse diagram of the Tamari lattice is isomorphic to the vertex-edge incidence graph of an associahedron. The number of elements in a Tamari lattice for a sequence of n + 1 objects is the n-th Catalan number (Loday 2004). The combinatorics of the Tamari lattices was studied using Hopf algebras (Loday and Ronco 1998; Hivert et al. 2005; Foissy 2009).

Figure 4.2 shows the Tamari lattices for 6 steps. It is a 2D presentation.

Figure 4.3 shows the Tamari lattice for 8 steps.

The lattice shows all the possible paths and their interrelationship.

It may be considered as a 3D presentation.

The connection of such paths with dual graded graphs, DGG, was analyzed by Josuat-Vergès (2010).

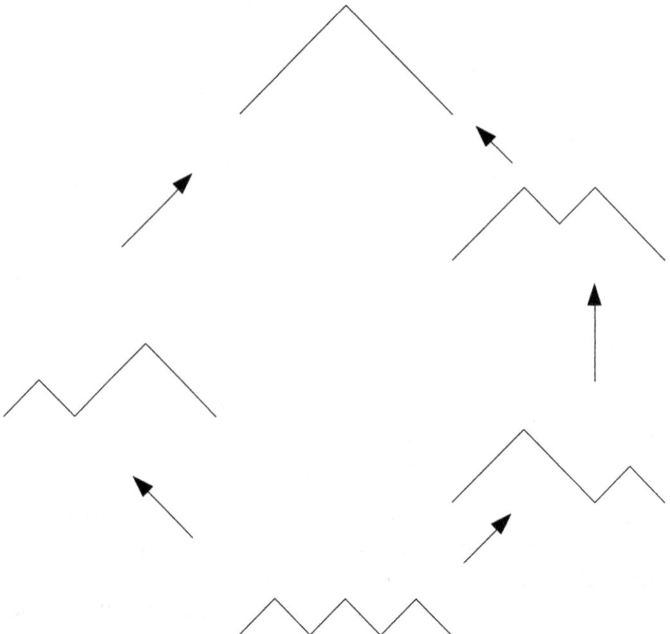

Fig. 4.2 Tamari lattice for 6 steps

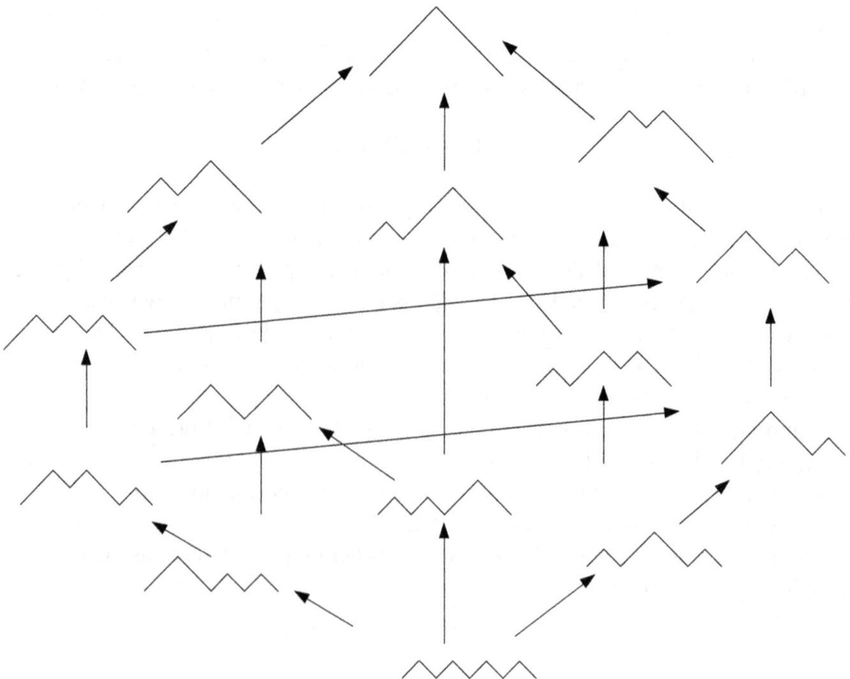

Fig. 4.3 Tamari lattice for 8 steps

4.1.2 *Growth Diagrams*

In many types of systems, atomic and molecular, biological, and sociological one wants to change global states by performing local actions. Growth diagrams are useful in organizing such local actions.

This refers to a procedure of how to move from the local level through the intermediate levels to the global level with respect to general properties and states. The importance of the level structure lies in the possibility of manipulating the systems level-wise in order to achieve a desired global goal or state.

The stages of a procedure, the devices of an installation or the compounds of a mixture to be separated may be denoted by integers 1, 2, 3 and so on. A succession of such integers is called a word or a permutation. According to specific rules this integers may be grouped in series or in parallel. The results of groupings are called partitions. Confronted with specific words or permutations the system answer is specific partitions. To a specified order of operations the installation answer is a specific architecture (Rey 2007).

A word is a request or a question and the smart system self-evolve according to specific rules to accommodate this request or answer this question. For a micro-reactors system this means to change the interconnections of modules and the timing of their activation.

Permutations may be associated to specific design of experiments in which for any experiment just one factor is modified. Experiments may be assigned to rows and factors to columns in the matrix of design associated to permutation.

There is a well-studied correspondence between words in the alphabet of positive integers and pairs consisting of a semistandard Young tableau and a standard Young tableau of the same shape called the Robinson-Schensted-Knuth (RSK) correspondence (Fomin 1992, 1994; Stanley 1997, 1999).

Given a word w, the RSK correspondence maps w to a pair of tableaux, (P, Q), via a row insertion algorithm consisting of inserting a positive integer into the tableaux.

The algorithm for inserting positive k into a row of a semistandard tableau is defined as follows. If k is grater than or equal to all entries in the row, add a box labeled k to the end of the row. Otherwise, find the first y in the row with $y > k$. Replace y with k in this box and proceed to insert y into the next row. To insert k into the semistandard tableau P, we start by inserting k into the first row of P. To create the insertion tableau of a word $w = w_1w_2 \ldots w_r$ we first insert w_1 into the empty tableau, insert w_2 into the result of the previous insertion, insert w_3 into the result of the previous insertion and so on until we have inserted all the letters of w. Denote the resulting insertion tableau by P(w). It is always a semistandard tableau.

To obtain a standard Young tableau from w, we define the recording tableau, Q (w), by labeling the box of $P(w_1w_2 \ldots w_s)/P(w_1w_2 \ldots w_{s-1})$ by s.

Consider for example w = 14253 (Patrias and Pylyavskyy 2014, 2015).

This word has insertion and recording tableaux shown in Fig. 4.4.

Fig. 4.4 (P, Q) table for w = 14253

P(w)=

1	2	3
4	5	

Q(w)=

1	2	4
3	5	

For $w = w_1w_2 \ldots w_k$ a permutation, we can obtain (P(w), Q(w)) from w by using growth diagrams (Fomin 1992, 1994).

Let us recall the classical Fomin's local rules or growth diagrams, which in particular can describe the RSK correspondence.

First we create a $k \times k$ array with an x in the w_i-th square from the bottom of column i. For example, if w = 14253, we have the array shown in Fig. 4.5 (Patrias 2016).

Figure 4.5 shows a table associated to permutation w.

This codifies the request or the experiment to be done.

This table may be considered as a DOE matrix.

We label the corners of each square with a partition. We begin by labeling all corners along the bottom row and left side of the diagram with the empty shape, Ø.

To complete the labeling of the corners, let us suppose the corners μ, λ, and ν are labeled, where μ, λ, and ν are as in Fig. 4.6 (Krattenthaler 2006).

Figure 4.6 shows the labeling of the corners.

We use the word square when referring to a square in the growth diagram and the word box or cell when referring to a box in a partition.

We are interested in growth diagrams which obey the following forward local rules:

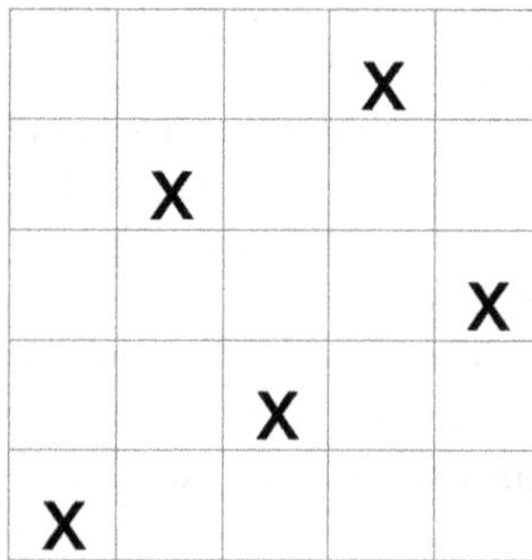

Fig. 4.5 Table associated to permutation

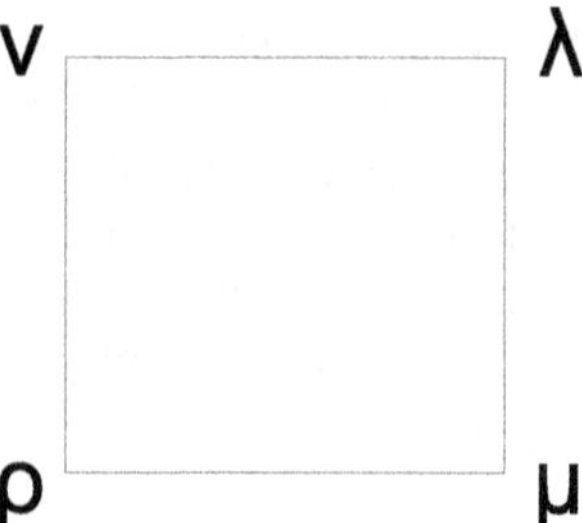

Fig. 4.6 Labeling of the corners

(F1) If $\rho = \mu = \nu$ and if there is no X in the cell then $\lambda = \rho$

(F2) If $\rho = \mu \neq \nu$ then $\lambda = \nu$

(F3) If $\rho = \nu \neq \mu$ then $\lambda = \mu$

(F4) If ρ, μ, ν are pairwise different then $\lambda = \mu \cup \nu$

(F5) If $\rho \neq \mu = \nu$ then λ is formed by adding a square to the (k + 1) st row of $\mu = \nu$, given that $\mu = \nu$, and ρ differ in the k-th row.

(F6) If $\rho = \mu = \nu$ and if there is an X in the cell then λ is formed by adding a square to the first row of $\rho = \mu = \nu$.

It is possible to work in another direction that is given λ, μ, ν one can reconstruct ρ and the filling of the cell.

The corresponding backward local rules are:

(B1) If $\lambda = \mu = \nu$ then $\rho = \lambda$

(B2) If $\lambda = \mu \neq \nu$ then $\rho = \nu$

(B3) If $\lambda = \nu \neq \mu$ then $\rho = \mu$

(B4) If λ, μ,ν are pairwise different then $\rho = \mu \cap \nu$

(B5) If $\lambda \neq \mu = \nu$ then ρ is formed by deleting a square from the (k-1) st row of $\mu = \nu$ given that $\mu = \nu$ and ρ differ in the k-th row $k \geq 2$

(B6) If $\lambda \neq \mu = \nu$ and if λ and $\mu = \nu$ differ in the first row, then $\rho = \mu = \nu$

In case (B6) the cell is filled with an X. Otherwise the cell is filled with 0.

Following such rules, there is a unique way to label the corners of the diagram. The resulting array is called the growth diagram of w, and rules (F1)–(F6) are called forward growth rules.

Figure 4.7 shows the growth diagram for the word w = 14253.

The cells from the rules are represented here by integers (Patrias 2016).

From the chains of partitions on the rightmost edge and uppermost edge of the growth diagram we can read the insertion tableau P, and recording tableau Q, respectively (see Fig. 4.4).

For w = 2413, the pair (P, Q) is shown in Fig. 4.8.

The RSK algorithm gives a bijection between permutations of n and a pair (P,Q) of standard Young tableaux of the same shape with n boxes.

For the pair (P, Q) shown in Fig. 4.8, the Young tableaux may be illustrated using the Young lattice as shown in Figs. 4.9 and 4.10 (Patrias and Pylyavskyy 2014, 2015).

Standard Young tableaux are represented as paths from ∅ in Young's lattice.

Through the language of growth diagrams, any explicit pairing of up-down and down-up paths that shows DU − UD = I in any dual graded graph, DGG, determines a RSK-like insertion algorithm (Appendix A).

Figure 4.9 illustrates the P-path with the help of thick interconnections.

The integers are represented here by cells.

For instance, the elements of the top row of this tree are: 1111; 211; 22; 31; 4.

The elements of the next row in the tree are: 111; 21; 3 and so on.

Figure 4.10 illustrate the Q-path with the help of thick interconnections..

We observe that it is a relation between the mixing as studied in classical diffusion theory and the insertions and growth as studied in lattice theory.

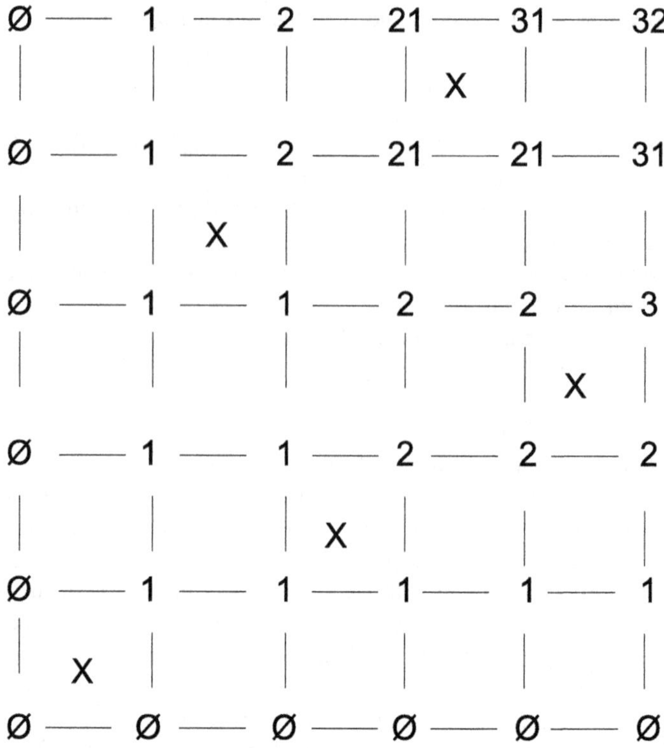

Fig. 4.7 Growth diagram

Fig. 4.8 (P, Q) table for w = 2413

P(w)=

1	3
2	4

Q(w)=

1	2
3	4

The forward Kolmogorov eq., the residence time distribution RTD, the U-graph in DGG is related to the forward evolution.

Complementing this is the backward Kolmogorov eq., the traveling time distribution TTD, the D-graphs in DGG all are related to the backward evolution.

The challenge is to find and impose specific rules of transition between such complementary ways. A process with specified transitions between forward and backward evolutions would be of practical interest. It offers the possibility of manipulating the systems level-wise in order to achieve a desired goal or state.

Since complementary ways and descriptions may be in disagreement, the conceptual and physical integration or coordination of complementary descriptions requires a higher level description, a new dimension that emerges as a new hierarchical level of complexity, as a confluence center.

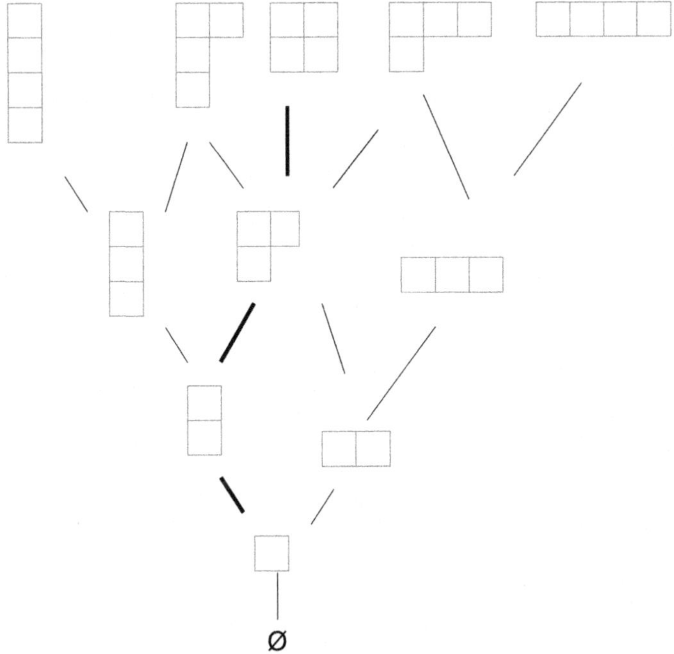

Fig. 4.9 P-path

The swing between forward and backward steps may be the source of innovative high- dimensional mixing processes. Correlating the forward and backward processes will allow us to perform mixing, heat and mass transfer, that could not be done by only one of these processes. This is the proposed strategy for high-dimensional mixing.

4.2 Restricted Processes

4.2.1 Diagonal Rectangulations

A rectangulation is a way of decomposing a square into rectangles.

A diagonal rectangulation of size n is a n × n square divided into n rectangles such that the interior of each rectangle intersects the diagonal of the square with negative slope.

Figure 4.11 shows an example of diagonal rectangulation.

It may be of interest to study the restricted transport from the upper-left corner (0; n), to the lower-right corner at (n; 0). This is a percolation in a rectangular domain.

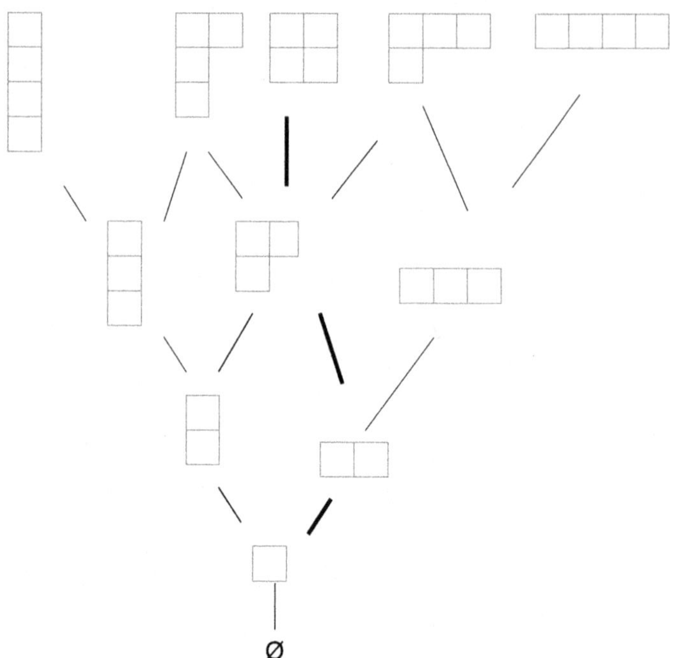

Fig. 4.10 Q-path

Fig. 4.11 Diagonal rectangulation

The Hopf Algebra of diagonal rectangulations was studied by Law and Reading (2012) and by Meehan (2017).

Let dRecn denote the set of diagonal rectangulations of size n.

Defining the product and coproduct as described below, we obtain the Hopf algebra of diagonal rectangulations.

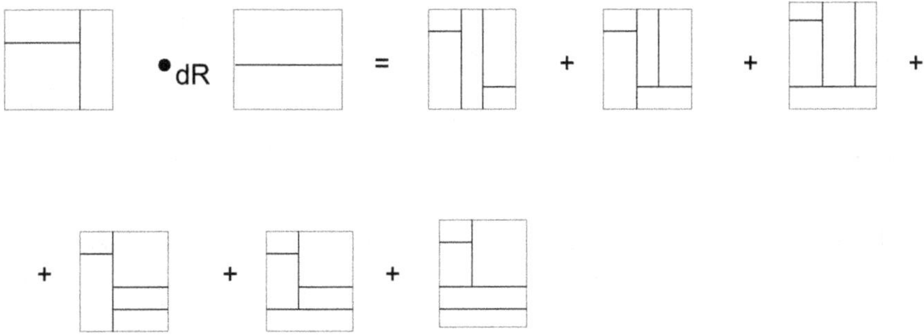

Fig. 4.12 Product

Let C denote a collection of line segments and points contained in a square. We say that C is a partial diagonal rectangulation if there exists a diagonal rectangulation D such that each line segment of D contains either a line segment or point of C. We call such a diagonal rectangulation D a completion of C.

Figure 4.12 shows the product of two diagonal rectangulations in the Hopf algebra of diagonal rectangulations.

Let D1 > dRecp and D2 > dRecq. To find the product of D1 and D2, denoted D1 • $_{dR}$D2, we begin with a p + q unit square with lower-left vertex at (0; 0). Place D1 and D2 in this square so that the upper-left corner of D1 is at (0; n), the lower-right corner of D2 is at (n; 0), and the lower-right corner of D1 and upper-left corner of D2 are at (p; q).

Remove the bottom and right side of D1 and the top and left side of D2 from this diagram.

Let C denote the union of the collection of remaining line segments, the boundary square, and V = {(p; q)}. Then D1 • $_{dR}$D2 is the sum of all completions of C.

The product in the Hopf algebra of diagonal rectangulations can be described as the sum of the elements of an interval in a lattice.

We say that P is a path in a diagonal rectangulation D if P joins the upper-left vertex of the boundary square to the lower-right vertex of the boundary square and consists of down and right steps along edges of D.

For each such path, let $R_l(P)$ denote the union of the boundary of the square and edges of D below P, and let $R_u(P)$ denote the union of the boundary of the square and the edges of D above P. The coproduct of the Hopf algebra of diagonal rectangulations is:

$$\Delta_{dR}(D) = \sum\nolimits_{P} \left(\sum\nolimits_{c} R_l(P) \otimes \sum\nolimits_{c} R_u(P) \right) \tag{4.5}$$

Figure 4.13 illustrates the coproduct.

Figure 4.14 shows a polytope with diagonal rectangulations for different planes.

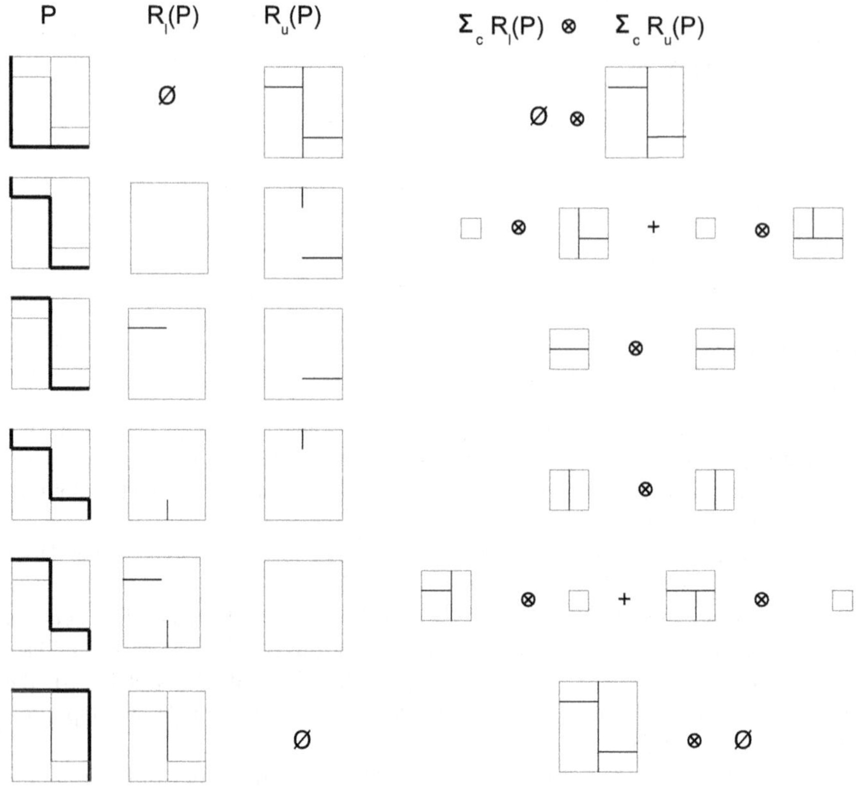

Fig. 4.13 Coproduct

Path in a diagonal rectangulation joins the upper-left vertex of the boundary square to the lower-right vertex of the boundary square and consists of down and right steps along edges. If the available space is 6D we are faced with a 6D rectangulation.

The trajectories may be extremely complex. Only few samples of rectangulation have been shown in Fig. 4.14.

With the thick hyper-lines, the 6D cube is presented as a "cube of cubes".

4.2.2 Processes Intensification

Chemical industries and process engineering are undergoing rapid changes facing the challenges of climate change, energy shortage and complexity advent. Process Intensification (PI) that leads to a smaller, less costly, cleaner, safer, higher productivity and more energy efficient technology was proposed as a new paradigm of process engineering (Stankiewicz and Moulijn 2002).

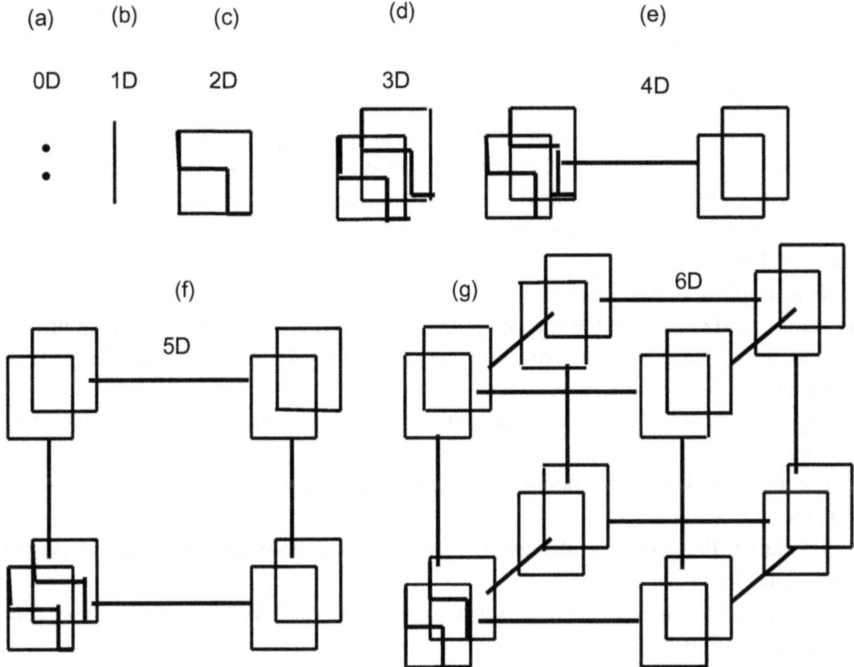

Fig. 4.14 Polytope 6D for diagonal rectangulation

Particularly, the innovative design of high yield processes or compact equipments has become one of the pressing industrial needs in recent years.

One of the routes to PI is the use of equipments with locally miniaturized structures, that is, micro or mini-channels because of their enhanced heat and mass transfer properties. Miniaturized process and energy equipments can be heat exchangers, chemical mixers or reactors, fuel cells (Ehrfeld 2007; Ehrfeld et al. 2000). Nevertheless, to obtain a comparable productivity with that of conventional equipment, a number of micro/mini-channels should be installed in parallel.

This so-called numbering-up process is the key issue for large-scale industrial applications of miniaturized devices (Coppens 2005; Kashid et al 2010).

It is a distinctive scale-up method. The numbering-down process is a scale-down method.

The fluid distribution uniformity among the parallel channels may play an important role on the global performance of multi-channel equipments.

This is particularly true when multi-scale ladder-type fluid networks are involved (Saber et al. 2010; Commenge et al. 2011; Luo 2013; Pistoresi et al 2015). In order to achieve uniform flow distribution among all parallel channels in the network, the essential step is the design of an optimized two-scale elementary Z-type ladder circuit. In this elementary fluidic circuit, the single inlet port and single outlet port are located on opposite sides of the bundle of parallel cross-channels, meaning that the flow direction is the same in the distributor and the collector pipes.

All cross-channels are assumed to have the same geometrical characteristics so that the passage-to-passage maldistribution may be considered as negligible. On the contrary, the improper design of fluid distributor and collector pipes configuration is the main cause of flow maldistribution among the parallel cross-channels.

Many studies have then been focused on how to improve the flow distribution uniformity of the elementary Z-type ladder circuit. A relatively uniform distribution may only be approached by making their hydraulic resistance much larger than that of the distributor and collector pipes if the latter have a uniform profile (Saber et al. 2009, 2010).

This implies that the distributor and collector pipes are large and hindering, which is clearly unfavorable for miniaturized devices. Instead of the uniform profile o rectangular shape of the distributor and collector, alternative shapes have been proposed.

Figure 4.15 shows the development of polytopic architecture for multi-scale transfer.

Figure 4.15 shows the increasing of dimension from 0D to 6D for multi-scale transfer.

To clarify the proposed design, Fig. 4.16 shows more details for polytope 4D multi-scale transfer.

Figure 4.16a illustrates the absence of transfer that is 0th scale or no scale available for transfer.

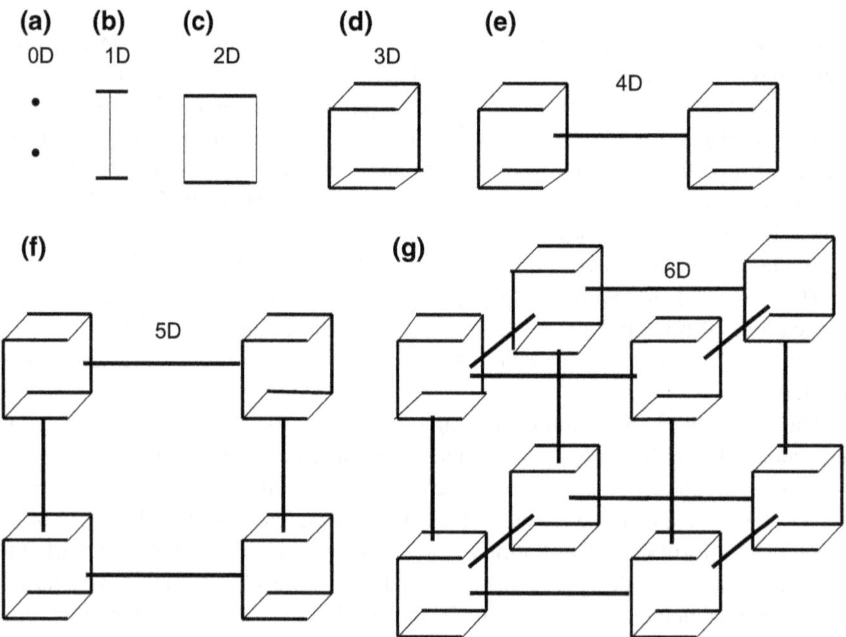

Fig. 4.15 Polytope 6D for multi-scale transfer

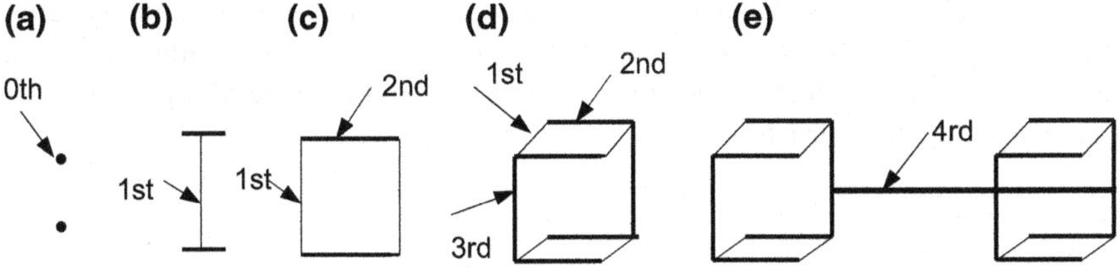

Fig. 4.16 Details for polytope 4D for multi-scale transfer

Figure 4.16b shows just one micro-channel and the 1st scale of transfer.

Figure 4.16c shows several micro-channels. They are in the same plane. Only two are indicated here but could be a plane network.

Figure 4.16d shows several parallel micro-plates. They are the top and bottom face of the cube. As before several plates may be included but the device is 3D. The 3rd scale of transfer is included here.

Figure 4.16e shows one micro-device. It results by interconnecting the 3rd scale network.

The thick hyper-line connecting the 2 cubes in 4D device represents 8 connections between corresponding vertices of the cubes. But this is only the ideal 4D situation.

Less than 8 connections need to be considered in practical situations.

Figure 4.17 shows details for polytopes 5D and 6D for multi-scale transfer

Figure 4.17a and b shows micro-devices resulting from 5D and 6D geometry.

They result by simplifying the interconnections specific to complete 5D and 6D geometry. Such geometries have been studied in parallel computer architecture (Seitz 1985; Ammon 1998).

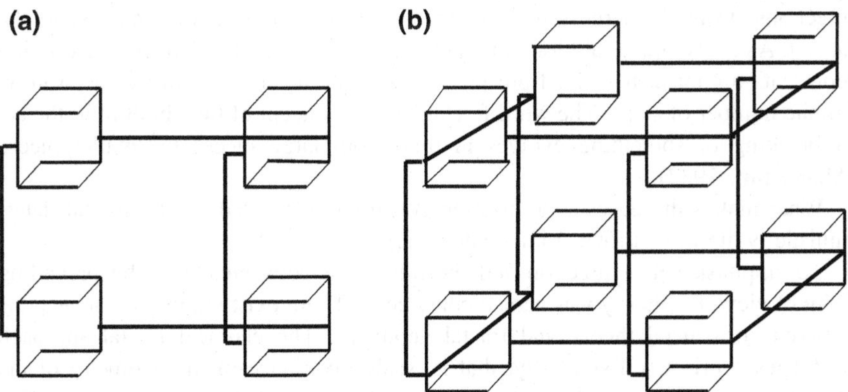

Fig. 4.17 Details for polytopes 5D and 6D for multi-scale transfer

One of the research challenges is to design real high-dimensional systems. Figure 4.17 illustrates the sort of design one is led to. This figure illustrates how heat pipes or heat conducting rods may be combined to move the heat into or out of a poorly conducting structure.

4.2.3 Fractal or Constructal

The way of building multi-scale structures from the largest to the smallest or from the smallest scales to the largest is typical of the fractal or constructal approach.

Various transfer structures existing in nature have been analyzed using the fractal geometry. The analysis has extended to the engineering, such as the research of heat and mass transfer problems (Mandelbrot 1982; Kearney 1999, Coppens 2004, 2005; Pence 2010).

Fractal is a self-similar structure with extended symmetry. This is a process from large to infinite small, which can approximately describe some geometric shapes in nature. If a sufficient number of self-similar iterative processes are carried out by computer, one can get a vivid tree shape. This is a descriptive and non-deterministic process from large to small. Compared with the fractal theory, the constructal theory describes a process from the finite small unit to the large structure, and the arrow in time is just the opposite to that of the fractal theory (Bejan 2000; Bejan and Tondeur 1998; Chen 2012).

A major task of the fractal theorists was to simulate some structures as accurately as possible. The constructal theory claims that it is able to predict the structures of objects and guide the design of engineering devices.

The fractal algorithms that are used to generate patterns that look like natural tree networks align themselves with the time direction favored but not entirely in physiology, from large to small. The postulated algorithm can be executed in both directions, from the largest scale to the smallest, and from the smallest to the largest. As a descriptive aid for natural phenomena, however, the fractal description represents a choice, namely, from the largest scale all the way to size zero in an infinite number of steps. The word fractal has the concept of time built in it: the act of breaking of something evolves in time from large pieces to smaller pieces (Mandelbrot 1982).

What makes the access-optimization construction labeled as constructal deterministic is the time arrow, from small to large.

To emphasize this direction and the difference between it and the prevailing point of view, the theory was named constructal. There exist important connections between constructal theory and fractal geometry. The classical fractal approach postulates a priori self-similarity, that is, scale invariance of the geometry of the objects, or invariance in the zoom direction. Constructal theory aims to establish relations between successive scales (scale covariance, not invariance) as a deterministic result of constrained optimization (Bejan and Tondeur 1998).

Consider for example the idea that large flows split into two smaller flows, each of which again splits into two still smaller flows, and so on, like streams in the delta of a river. We observe that the splitting may also run the other way that is merging, like creeks merging into rivers. Numerous physical systems contain both splitting and merging, for instance the blood system which splits from arteries to arterioles to capillaries in the tissue and then merges back to veins (Andersen 2013).

The concepts of life, design, and evolution were placed in physics by the so-called constructal law (Bejan 2000):

> For a finite-size flow system to persist in time (to live), its configuration must evolve in such a way that provides greater and greater access to the currents that flow through it.

According to the constructal law, a living system is one that has just two universal characteristics: It flows and it morphs freely toward configurations that allow all its currents to flow more easily over time. Life and evolution are considered by constructal theorists as physical phenomena, belonging to physics.

The significance of fractal concepts for like-alive systems has been the object of several studies (Weibel 1991; West et al. 1999).

To evaluate the role of fractal or constructal theories for living systems characterization we need to start from the understanding of life as a distinct level of reality.

To understand a living or like-alive system as we know it, especially the continuous evolution of stable complex forms, it has proven essential to distinguish two complementary types of models and controls (Pattee 1995, 2000). One type is a semiotic or symbolic model exerting upward control from a local isolated memory. This covers the constructal model and control.

The other type, is a dynamic model exerting downward control from a global network of coherent interactive components. This covers the fractal model and control.

The constructal approach explains how control can be inherited and provides a more or less efficient search process for discovering adaptive structures.

The fractal approach suggests how the many components constructed under semiotic control can be integrated in the course of development and coordinated into emergent functions.

According to this understanding of life-alike systems neither model, fractal or constructal, has much explanatory and predictability value without the other.

Each model alone can account for a limited level of self-organization. For example, biosystems study shows that copolymers can self-assemble more or less randomly, and by chance form autocatalytic cycles. Dynamics can also generate numerous complex autonomous patterns. But dynamics without an open-ended heritable memory or memory without dynamic coordination have very limited emergent and like-alive potentiality.

The life probably requires the coupling of both self-organizing processes, fractal and constructal. Living and artificial like-alive systems need semantic or semiotic closure (Pattee 1995).

The correlation of fractal and constructal ways involves a supplementary dimension that is, the polytopic architecture.

Since complementary descriptions may appear as basically incongruous and contradictory the integration or coordination of complementary descriptions requires a higher level description that appears as a new hierarchical level of complexity that is a new dimension. The more complex is a system the more descriptions and dimensions are necessary for comprehension, design fabrication and control.

The swing between constructing and deconstructing steps in 3D implies a supplementary dimension the 4D and so on, as illustrated in Figs. 4.16 and 4.17.

The fourth dimension of life, fractal geometry and allometric scaling of organisms has been discussed by West et al. (1999) from a different point of view.

Correlating the constructing and deconstructing processes will helps make things or run operations that could not be made or performed by only one of these processes in duality.

The claim of universality for any description of reality does not imply that such a description is complete, nor does inconsistency between two complementary descriptions imply that one of them does not apply to reality. This is valid for fractal and constructal models,

Fractal or constructal represents a Cartesian split antinomy that may block the development towards like-alive systems (Overton 2003, 2013). For like-alive systems, fractal and constructal should stay togther as necessary and complementary dualities, acting in synergy rather than conflicting, as inclusive rather than exclusive.

According to the adopted understanding of life, like-alive technological systems able to challenge complexity advent will be nor fractal and nor constructal but most likely polytopic.

References

Ammon, J.: Hypercube Connectivity Within ccNUMA Architectures. SGI Origin Team, Mountain View, CA (1998)

Andresen, B.: Book review. Design in nature. J. Non-Equil. Thermod. **38**, 287–290 (2013)

Bejan, A.: Shape and Structure, from Engineering to Nature. Cambridge University Press, Cambridge, UK (2000)

Bejan, A., Tondeur, D.: Equipartition, optimal allocation, and the constructal approach to predicting organization in nature. Revue Générale de Thermique **37**(3), 165–180 (1998)

Chen, L.: Progress in study on constructal theory and its applications. Sci. China Technol. Sci. **55** (3), 802–820 (2012)

Commenge, J.-M., Saber, M., Falk, L.: Methodology for multi-scale design of isothermal laminar flow networks. Chem. Eng. J. **173**, 541–551 (2011)

Coppens, M.O.: Nature inspired chemical engineering learning from the fractal geometry of nature in sustainable chemical engineering. In: Fractal Geometry and Applications: A Jubilee of Benoit Mandelbrot, vol **72**, pp 07–32. 24 Dec 2004

Coppens, M.O.: Scaling-up and-down in a nature-inspired way. Ind. Eng. Chem. Res. **44**(14), 5011–5019 (2005)

Ehrfeld, W.: Process intensification through microreaction technology. In: Stankiewicz, A., Moulijn, J.A. (eds.) Re-engineering the Chemical Process Plant-Process Intensification, pp. 155–175. Marcel Dekker Inc., NY (2007)

Ehrfeld, W., Hessel, V., Lowe, H.: Microreactors: New Technology for Modern Chemistry. Wiley-VCH, Weinheim (2000)

Foissy, L.: The infinitesimal Hopf algebra and the poset of planar forests. J. Algebraic Comb. **30**, 277–309 (2009)

Fomin, S.: Dual graphs and Schensted correspondences. In: Leroux, P., Reutenauer, C. (eds.) Series formelles et combinatoire algebrique, pp. 221–236. Montreal, LACIM, UQAM (1992)

Fomin, S.: Duality of graded graphs. J. Algebraic Combin. **3**, 357–404 (1994)

Geyer, W.: On Tamari lattices. Discret. Math. **133**(1–3), 99–122 (1994)

Hivert, F., Novelli, J.-C., Thibon, J.-Y.: The algebra of binary search trees. Theoret. Comput. Sci. **339**(1), 129–165 (2005)

Iordache, O.: Stability of Microheterogeneous Dispersed Systems, Ph D Thesis, Polytechnical Institute Bucharest (1974)

Iordache, O.: Model with two equations for transfer phenomena. Rev. Chimie (Bucharest) **27**, 871–875 (1976)

Iordache, O.: Implementing Polytope Projects for Smart Systems. Springer, Cham, Switzerland (2017)

Josuat-Vergès, M.: Stammering tableaux. 10 Jan 2016. arXiv preprint arXiv:1601.02212

Kac, M.: A stochastic model related to the telegrapher's equation. Rochy Mountain J. Math. **4**, 497–509 (1974)

Kashid, M.N., Gupta, A., Renken, A., Kiwi-Minsker, L.: Numbering-up and mass transfer studies of liquid–liquid two-phase microstructured reactors. Chem. Eng. J. **158**(2), 233–240 (2010)

Kearney, M.: Control of fluid dynamics with engineered fractals-adsorption process applications. Chem. Eng. Commun. **173**(1), 43–52 (1999)

Krattenthaler, C.: Growth diagrams, and increasing and decreasing chains in fillings of Ferrers shapes. Adv. Appl. Math. **37**, 404–431 (2006)

Law, S., Reading, N.: The Hopf algebra of diagonal rectangulations. J. Comb. Theory Ser. A **119** (3), 788–824 (2012)

Loday, J.L.: Realization of the Stasheff polytope. Arch. Math. **83**(3), 267–278 (2004)

Loday, J.L., Ronco, M.O.: Hopf algebra and the planar binary trees. Adv. Math. **139**(2), 293–309 (1998)

Luo, L. (ed.): Heat and Mass Transfer Intensification and Shape Optimization: A Multi-scale Approach. Springer Science & Business Media, London (2013)

Mandelbrot, B.: The Fractal Geometry of Nature. W.H. Freeman, San Francisco (1982)

Meehan, E.: Posets and Hopf Algebras of Rectangulations. Ph D Thesis. North Carolina State University (2017)

Overton, W.F.: Understanding, explanation, and reductionism: finding a cure for Cartesian anxiety. In: Reductionism and the Development of Knowledge, pp. 39–62. Psychology Press (2003)

Overton, W.F.: Relationism and relational developmental systems: a paradigm for developmental science in the post-Cartesian era. Adv. Child Dev. Behav. **44**, 21–64 (2013)

Patrias, R: Combinatorial constructions motivated by K-theory of Grassmannians, PhD Thesis, Univ of Minnesota (2016)

Patrias, R., Pylyavskyy, P.: Dual filtered graphs (2014) arXiv preprint arXiv:1410.7683

Patrias, R., Pylyavskyy, P.: Dual filltered graphs extended abstract DMTCS Proc. FPSAC'15, pp 1–12 (2015)

Pattee, H.H.: Evolving self-reference: matter, symbols, and semantic closure. Commun. Cognit. Artif. Intell. **12**(1–2), 9–28 (1995)

Pattee, H.H.: Causation, control and the evolution of complexity. In: Anderson, P.B., et al. (eds.) Downward Causation. Aarhus University Press, Aarhus, Denmark (2000)

Pence, D.: The simplicity of fractal-like flow networks for effective heat and mass transport. Exp. Therm. Fluid. Sci. **34**, 474–486 (2010)

Pistoresi, C., Fan, Y., Luo, L.: Numerical study on the improvement of flow distribution uniformity among parallel mini-channels. Chem. Eng. Process. **95**, 63–71 (2015)

Rey, M.: Algebraic constructions on set partitions. In: Proceedings of 19th International Conference on Formal Power Series & Algebraic Combinatorics, Nankai University, Tianjin, China, 2–6 July 2007

Saber, M., Commenge, J.M., Falk, L.: Rapid design of channel multi-scale networks with minimum flow maldistribution. Chem. Eng. Process. **48**, 723–733 (2009)

Saber, M., Commenge, J.-M., Falk, L.: Microreactor numbering-up in multi-scale networks for industrial-scale applications: Impact of flow maldistribution on the reactor performances. Chem. Eng. Sci. **65**, 372–379 (2010)

Seitz, C.L.: The cosmic cube. Commun. ACM **28**(1), 22–33 (1985)

Stankiewicz, A., Moulijn, J.A.: Process intensification. Ind. Eng. Chem. Res. **41**(8), 1920–1924 (2002)

Stanley, R.P: Enumerative Combinatorics, vol. 1, Cambridge Studies in Advanced Mathematics, vol. 49, Cambridge University Press, Cambridge (1997)

Stanley, R.P: Enumerative Combinatorics, vol. 2, Cambridge Studies in Advanced Mathematics, vol. 62, Cambridge University Press, Cambridge (1999)

Weibel, E.R.: Fractal geometry: a design principle for living organisms. Am. J. Physiol. **261**, 361–369 (1991)

West, G.B., Brown, J.H., Enquist, B.J.: The fourth dimension of life: fractal geometry and allometric scaling of organisms. Science **284**(5420), 1677–1679 (1999)

Chapter 5
Assimilation and Accommodation

5.1 Developmental Stages

5.1.1 Cognitive Development

Information—processing theories have yielded measures of cognitive complexity, which can be considered a major organizing theme in cognitive development (Halford and Andrews 2010).

A dominant early approach has been the Piagetian and neo-Piagetian models, which sought to explain the course of cognitive development, in terms of the growth of information processing capability.

Cognitive structures are patterns of physical or mental actions that underlie specific acts of intelligence and correspond to the stages of development (Piaget 1970, 1971).

According to Piaget, there are four cognitive development stages:

- sensory-motor
- preoperational
- concrete operational
- formal.

Figure 5.1 shows the developmental stages hierarchy.

This is a 1D presentation. If formal stage goes back to modify sensory motor stage we will be faced with a 2D presentation of the development.

It was observed that restriction of cognitive capability to the formal stage may correspond to systems stagnation and possible failure for evolving automata (Bringsjord et al. 2010).

This refers to automata that have a code or protocol that recommend some actions for situations requiring a different code.

Growing complexity impose to look forward for creativity, self-evolvability and smartness for automata (Iordache 2012).

O. Iordache, *Advanced Polytopic Projects*, Lecture Notes in Intelligent Transportation and Infrastructure,
https://doi.org/10.1007/978-3-030-01243-4_5

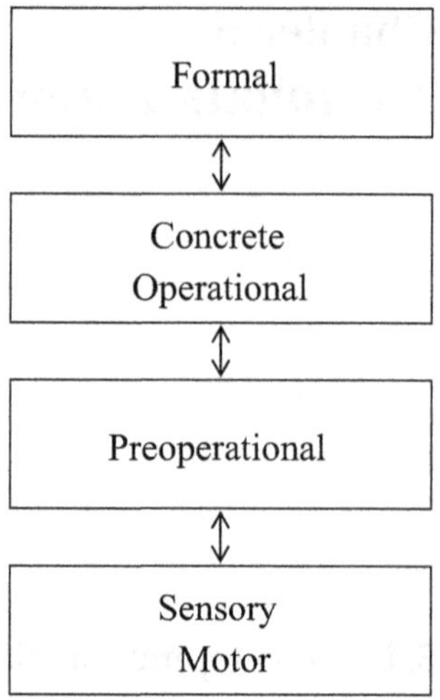

Fig. 5.1 Developmental stages hierarchy

Piaget's epistemology opens the road for cognition beyond the fourth stage. Piaget initiated the study of post-formal stages, in which agents are able to operate over logical systems. This refers to meta-processing of logics and formal theories expressed in those logics. Elaboration and modification of axiomatic schemes may be considered as surpassing the formal stage and are to formal schemes what the latter are to concrete operations (Piaget 1973; Piaget and Garcia 1989).

The post formal stages appeared as possible candidates for the so-called 5th cognitive development stage (Bringsjord et al. 2010). They are comparable to the formal framework in which post-formal reasoning involves the "Self".

Figure 5.2 shows a 4D polytopic presentation of the cognitive developmental stages.

The initial four stages of Piaget, associated to S, K1, K2 and K3, have been completed in Fig. 5.2 by the self-evolvability stages. This allows describing systems able to self-evolve by internal structures modification.

There are four stages on the front face of the polytope. The notations are:

- S-Sensory Motor
- K1-Preoperational
- K2-Operational
- K3-Formal.

The development is represented clockwise.

Piaget considered that the sensorimotor stage differed from the later stages in that the former was devoid of symbolic representation.

The central stage the "Self", may ensure the cooperation and redistribution of the four stages on another face of the polytope, with another starting stage.

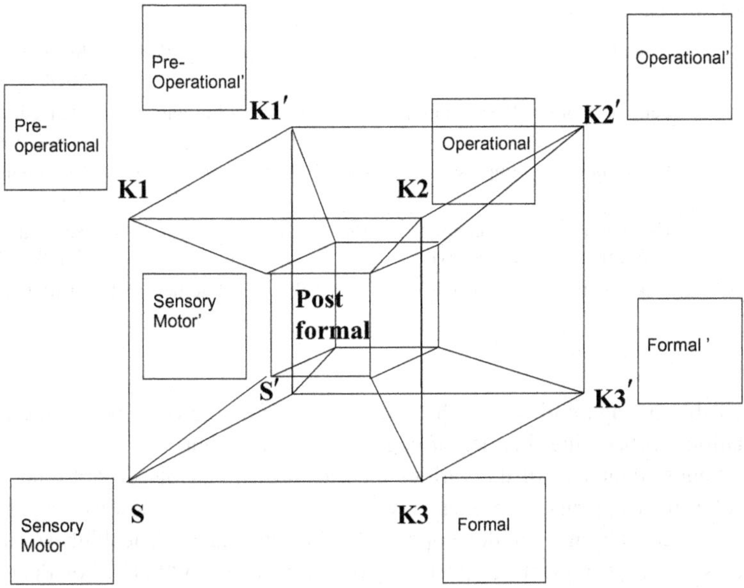

Fig. 5.2 Polytope 4D for development stages

Any stage embeds the previous ones. After one cycle an augmented reality may support a new cycle of development. The post-formal stage appears as a cognitive exemplar of the "Self".

As shown in Fig. 5.2 two ways should be considered for development.

This means that after the direct or constructive way S → K1 → K2 → K3 we need to look at the reverse or deconstructive way K3′ → K2′ → K1′ → S′.

We can identify:

- S′-Sensory Motor'
- K1′-Preoperational'
- K2′-Operational'
- K3′-Formal'.

The new stages are modified or even reversed.

Using the developments of the direct way, it is possible to modify results, reconsider cognitive capabilities for the reverse way. The swinging from direct to modified developmental stages mediated by the "Self" may be the source of creativity in complex problem solving or technology development.

That is because the boundaries where creative research stand out, the old information is re-evaluated and new information is created, consist of coexisting tendencies. Integration and differentiation coexists and the metastable coordination dynamics emerges as the specific blend of integration and differentiation tendencies.

Table 5.1 summarizes the information about development stages for different authors.

Table 5.1 Development stages

Author	Stage 0	Stage 1	Stage 2	Stage 3	Stage 4 "Self"
Piaget (1971)	Sensory-motor	Pre-operations	Concrete operations	Formal	Dialectical operations
Loevinger (1976)	Pre-social	Impulsive	Opportunistic	Conformist	Conscient Autonomous
Selman (1980)	Undifferentiate Egocentric	Differentiate Subjective	Reciprocal	Mutual	Societal Symbolic
Kegan (1979)	Incorporative	Impulsive	Imperial	Interpersonal	Institutional

The Table 5.1 again illustrates the underlying common ground between different information—processing theories of cognitive development.

The stages sequences of the various theories are similar but do not coincide.

Development appears as broken down in five stages. The lowest stage 0 corresponds to the absence of development or starting stage. The highest stage 4 represents the development perspective and is associated to "Self". For Piaget this stage corresponds to post-formal dialectical operations.

We refer more in detail to the theory of Kegan (1979).

Kegan uses the term constructive-developmental instead of cognitive-developmental and defines constructive development as the union of two separate ideas: constructivism and development.

Constructivism refers to the fact that cognitive systems constitute or construct reality.

Development refers to the fact that organic systems evolve through periods according to regular principles of stability and change.

Kegan pointed out that the constructive developmental framework encompasses:

- adaptive relationship of organism and environment
- ego's dialectic of self and other
- truth-creating relationship of subject and object.

Kegan considered all three systems of adaptation, ego-development, and truth to be different foci of a single process; that of meaning-constitutive evolution. The "Self" creates meaning as it evolves, and each developmental stage indicates a whole new way of making meaning.

5.1.2 *Dynamic Skill Theory*

The theory of Fischer and coworkers (Fischer 1980, 2008; Fischer and Rose 1996) was based on cognitive skill theory, where skill refers to control over sources of variation in a person's own behavior.

There are four major developmental stages or tiers:

- Reflexes
- Sensorimotor actions
- Representational
- Abstract.

These can be compared to Piagetian stages. The stages are similar but do not coincide.

Within each tier there is a recurring cycle of four sub-levels:

- set
- mapping (of sets)
- system (composition of mappings)
- system of systems.

This decomposition in set, mapping, system and system of systems has a general validity, for different tears, and for different levels in complex systems studies.

The highest level of one tier is shared with the lowest level of the next, and represents a transition between tiers. This sharing of levels is not general for complexity studies.

A set is a source of variation over which some cognitive process exercises control.

A set implies correspondences between events or objects and actions since the thing is always included with the behavior in the definition of a set (Fischer 1980).

Complexity increases within a tier in a manner that bears a correspondence to McLaughlin's (1963) complexity scale. The four sub-levels within a tier comprise a set, then a mapping between two sets, then a system which is a mapping between two mappings of sets and is equivalent to four sets, then a mapping between two systems, that is a system of systems, which comprises eight sets.

This corresponds to McLaughlin's (1963) four sub-levels defined as $2^0 = 1$, $2^1 = 2$, $2^2 = 4$ and $2^3 = 8$ concepts considered simultaneously. Observe that this is analogous to doubling, as complexity increases, in the general polytopic framework.

This illustrates the essential common ground between different theories of cognitive development and justifies the polytopic projects attempt to unify and standardize the research discovery, design and control methods.

Fischer also has an extensive consideration of transformation rules for creating the transition from one level to another. The first rule is inter-coordination, which is a process of combining skills at one level to produce a skill at the next level.

Fischer (1980) formulation has been retained in essence in his later work, but there have been two major developments of the model. One has been to link the model to dynamic growth functions while the other has been to link it to spurts in brain growth, that is to the biological support (Fischer and Rose 1996; Fischer and Bidell 2006).

Firstly, it is proposed that the major reorganizations between sub-levels, as outlined above, correspond to growth spurts or other discontinuities in brain growth. Furthermore, these dynamic changes can occur concurrently in many independent systems, which might be localized in different regions of the brain. It is also proposed that each new level is marked by a new behavioral control system, which is supported by a new kind of neural network (Fischer and Rose 1996). Fischer et al. (2008) propose that knowledge is built from repeated reconstructions that move towards higher complexity and abstraction, though with many reversals and recoveries.

Fischer also proposes that the recurring cycles of development that occur in each tier are supported by observations of brain growth (Fischer and Bidell 2006).

The implications of the recent developments of Fischer theory are profound (Halford and Andrews 2010).

They offer a resolution of the anomaly, crucial to understanding development, that there is both variability and consistency in cognitive development. Dynamic systems produce variability from a relatively small set of common processes. The application of dynamic systems theory to the database provided by cognitive developmental stage theory is one of the main achievements of Fischer's theory.

The four tiers in the Fischer theory are denoted as follows:

- Reflexes-Ref
- Actions-Act
- Representations-Rep
- Abstractions-Abs.

The four sub-levels associated to any tier, and the modified four sub-levels are denoted:

0-set
1-mapping
2-systems
3-systemSystem of systems
0'-modified set
1'-modified mapping
2'-modified systems
3'-modified system of systems.

Figure 5.3 shows the polytope 5D for dynamic skill development.

Using the thick hyper-lines, the 5D cube is shown as "square of cubes".

The main tiers are denoted: Ref, Act, Rep and Abs.

If a fifth tier is activated this is referred in dynamic skill theory as Principles.

This allows to 5D frames to evolve to 6D frames by modifying Principles.

Figure 5.4 shows the polytope 6D for dynamic skill development.

With the thick hyper-lines, the 6D cube is shown as "cube of cubes".

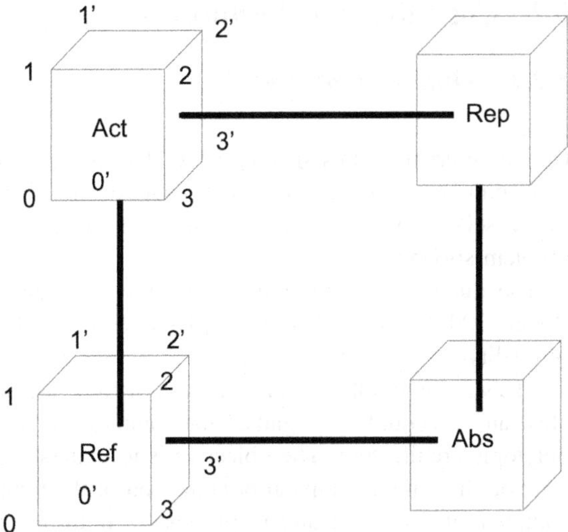

Fig. 5.3 Polytope 5D for dynamic skill development

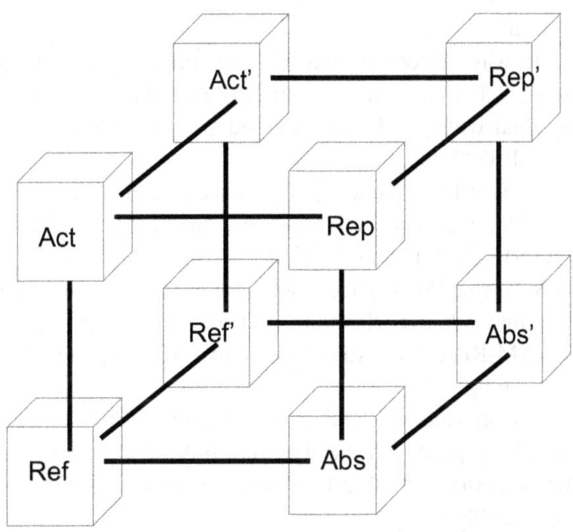

Fig. 5.4 Polytope 6D for dynamic skill development

The main tiers Ref, Act, Rep, and Abs are modified to the main tiers Ref', Act', Rep', and Abs' to account for the modified Principles. Principles are not illustrated in Fig. 5.4 but they play the central role in transition from 5D to 6D. In fact exactly such centers allow high-dimensional evolution. Following the increase of dimension, the associated center is no more necessary. But for the next dimensional increase a new center is required.

5.2 Cognitive Architectures

5.2.1 Duality Features

Cognitive architectures specify the underlying structure for intelligent systems.

Cognitive architectures play a vital role in providing blueprints for building future self-evolvable smart systems supporting a broad range of capabilities in problem solving.

The duality for cognitive architecture is related to dual-process theory, DPT (Evans 2008) dual graded graphs, DGG (Fomin 1992, 1994) and to other theories of duality.

The starting point is that numerous cognitive architecture are similar or relatively close and we could try to gather that common elements in a general architecture the polytopic architecture. The objective is to understand how the cognitive processes develop, how the dynamical organization of the brain structure can be viewed in a computational context and to propose a multi-purposes electronic device able to take advantages of this architecture.

Table 5.2 refers to several examples of cognitive architectures artificial or natural.

It was observed that the cognitive architectures show multiple levels and sub-levels that is, they are multi-level architectures that they may be characterized by dual ways of divergence and convergence processes organized by a center the so-called "Self".

Some of the names of cognitive architectures are abbreviations as for instance: CDZ-Convergence Divergence Zone (Damasio 1989, 1994, Meyer and Damasio (2009), DLA-Dynamic Link Architecture, GNW-General Neuronal Workspace, GW-Global Workspace, HRR-Holographic Reduced Representations, LAMINART-Laminar Adaptive Resonance Theory, LIDA-Learning Intelligent Distribution Agent, RAAM-Recursive Auto Associative Memory, STAR-Structured Tensor Analogical Reasoning.

To provide a foundation for dynamical laws of cortical phenomena von der Malsburg proposed the dynamic link architecture DLA (von der Malsburg 2002). DLA theories involve first orders, second orders and third orders networks (Lücke et al. 2008).

Aspects of duality parallel to divergence and convergence as recorded in Table 5.2 have been discussed in cognitive sciences in the so-called dual process theories.

Dual processing accounts of cognition have been developed in a range of areas including learning, attention, reasoning, decision making and social cognition (Evans 2008).

In spite of a considerable degree of independence in the formulation of these accounts, they include a number of striking similarities. Processes that are rapid, automatic and effortless on the one hand are contrasted with those that are slow,

Table 5.2 Dual ways for cognitive architectures

Cognitive Architecture	Convergence	Divergence	References
ACT-R	Symbolic	Sub-symbolic Dynamic	Anderson and Lebiere (2014)
AURA	Concatenation	Superimposing	Austin (1995) Kustrin and Austin (1999)
CDZ	Convergence Unconscious	Divergence Conscious	Meyer and Damasio (2009)
CLARION	Implicit Tacit	Explicit	Sun (2007) Sun et al. (2001, 2005)
CogAff	Reactive	Reflective Deliberative	Sloman (2001)
DLA	Binding	Debinding	von der Malsburg (1994, 2002)
DORA	Reflexive	Reflective	Doumas et al. (2008)
DUAL	Symbolic	Connectionist	Nestor and Kokinov (2004)
GNW	Sensing	Acting	Dehaene et al. (2003)
GW	Unconscious	Conscious	Baars (1988, 2002)
HRR	Binding	Superposition	Plate (1995)
HTM	Convergence	Divergence	Hawkins and George (2009)
ICARUS	Perceptual	Intentional	Langley and Choi (2006)
INFERNET	Reflexive	Reflective	Sougne (1999)
K-SETS	Perceptual	Intentional	Freeman (2000)
LAMINART	Excitatory Ascending	Inhibitory Descending	Grossberg (1999, 2000, 2012)
LIDA	Unconscious	Conscious	Franklin (2006) Shanahan (2006, 2008)
LISA	Recipient Reflexive	Driver Reflective	Hummel and Holyak (1997)
RAAM	Compressor Encoding	Reconstructor Decoding	Pollack (1990)
SHRUTI	Associative	Rational	Shastri and Ajjanagadde (1993)
STAR	Associative	Relational Deductive	Halford et al. (1998)
Tripartite	Autonomous	Reflective Algorithmic	Stanovich (1999, 2010)

sequential and controlled on the other. These theories typically characterize the two processes as independent sources of control for behavior that may come into conflict and competition.

A number of theorists have mapped these dual processes on to two distinct cognitive systems. These systems have been given various names including experiential-rational, heuristic-analytic, heuristic-systematic, implicit-explicit, associative and rule-based and the neutral System 1 and System 2 (Stanovich 1999). The characteristics attributed to these underlying systems show quite a large degree of consensus across theories and domains of application. For example, System 1 using this as a generic label for the fast, automatic system, is often described as evolutionarily old, shared with other animals and independent of individual differences in general intelligence, whereas System 2 is by contrast evolutionarily recent, uniquely human and related to heritable differences in intelligence and working memory capacity.

Given this degree of consensus in dual system theories developed across different cognitive domains, it is tempting to conclude that the brain must indeed contain two systems broadly as described. System 1 and System 2 may be associated to divergent and convergent systems respectively in the cognitive architectures. The cooperation of the dual ways by the "Self" is mandatory.

Table 5.3 refers to dual systems theory. The enumerated properties refer to processes, methods, contents, architectures and evolution aspects.

Overton and Ricco (2010) observed that DPT may find its roots in Kant' proposed structure of mind. Kant's "forms of intuition" (System (1) which were understood to interact directly with the sensible world provide input for "categories of understanding" (System (2) yielding an object world that evidences both necessity and universality.

Table 5.3 Dual systems

Characteristics	System 1	System 2	References
Processes	Associative	Deductive Rule-based	Sloman (1996)
Processes	Unconscious	Conscious	Carruthers (2006)
Processes	Automatic	Controlled	Schneider and Schiffrin (1977)
Processes	Intuitive Heuristic	Analytic Formal	Hammond (1996) Evans and Over (1996)
Processes	Reflexive	Reflective	Lieberman et al. (2002)
Method	Implicit Tacit	Explicit	Reber (1993) Evans and Over (1996)
Method	Fast operating	Slow operating	Carruthers (2006)
Method	Experimental	Rational	Epstein (1994) Epstein et al. (1996)
Method	Non-Verbal Sensorial	Verbal Linguistic	Frankish (2010)
Content	Concrete	Abstract	Frankish (2010)
Content	Contextual	De-contextual	Frankish (2010)
Content	Specific	General	Frankish (2010)
Content	Spatial	Temporal	Wilson (2002)
Architecture	Parallel	Serial	Carruthers (2006)
Architecture	Modular	Cognitive Fluid Single system	Evans and Over (1996) Mithen (1996)
Evolution	Old	New	Frankish (2010)

Table 5.4 Dual Graded Graphs

Type	U-graph	D-graph	References
Binary rooted Trees	Lattice of binary trees	Bracket trees	Fomin (1992, 1994)
Lifted Binary Trees	Lifted binary trees	Binword	Fomin (1992, 1994)
n-Ary Trees	U-graph	D-graph	Fomin (1994)
Rooted Trees, RT	RT, U-graph	RT, D-graph	Hoffman (2003, 2008)
Catalan Trees CT	CT, U-graph	CT, D-Graph	Qing (2008)
Fibonnaci Graphs, FG	FG, U-graph	FG, D-graph	Fomin (1994)
Composition Trees, CT	CT, U-graph	CT, D-graph	Bjorner and Stanley (2005)
Pascal Graphs PG	PG, U-graph	PG, D-graph	Fomin (1994)
Partition Trees	U-graph	D-graph	Hoffman (2008)
Partitions	SYT-tree and Schensted graphs	Schensted graphs	Fomin (1994)
n-Core Graphs	U-graph	D-graph	Berg et al. (2011)
Shifted Shapes	U-graph	D-graph	Fomin (1994)
Twin Trees TT	TT, U-graph	TT, D-graph	Giraudo (2011)
Reflected Graphs, RG	RG, U-graph	RG, D-graph	Fomin (1994)
Spanning Subgraphs	U-graph	D-graph	Sloss (2005)
Permutation Trees, PT	PT, U-graph	PT, D-graph	Fomin (1994)

To characterize duality of cognitive architecture and processes we may use dual graded graphs DGG (Fomin 1992, 1994) (Appendix A). DGG represents a significant generalization of the concept of differential posets introduced by Stanley (1988).

The U-graph is associated to convergence and System 1 while the D-graph is associated to divergence and System 2.

Table 5.4 refers to dual graded graphs.

Notations are RT-rooted trees, CT-Catalan trees, FG-Fibonnaci Graphs, PG-Pascal Graphs and so on.

SYT denotes standard Young tableaux (Fomin 1994).

5.2.2 *Polytopic Architecture*

Baars' GW (global workspace) theory has inspired a variety of related consciousness models (Baars 1988, 2002). The central idea of GW theory is that conscious cognitive content is globally available for diverse cognitive processes including attention, evaluation, memory, and verbal report. The notion of global availability is suggested to explain the association of consciousness with integrative cognitive processes such as attention, decision making and action selection. Also, because global availability is necessarily limited to a single stream of content, GW theory may naturally account for the serial nature of conscious experience.

GW theory was originally described in terms of a blackboard architecture in which separate, quasi-independent processing modules interface with a centralized, globally available resource. This cognitive level of description is preserved in the computational models of Franklin (2006), who proposed a model consisting of a population of interacting software agents, and Shanahan (2006, 2008) whose model incorporates aspects of internal simulation supporting executive control and more recently spiking neurons.

A central global workspace, GW, constituted by long-range cortico-cortical connections, assimilates other processes according to their significance.

A neuronal implementation of a global workspace, GW, architecture, the so-called global neuronal workspace, GNW, was studied (Dehaene et al. 2003).

According to the GNW architecture, there are two distinct computational spaces in the brain (Dehaene et al. 2003). The first space consists of a collection of separate and anatomically confined processors, each of which is specialized in a particular function, for instance, visual motion processing. These functions are carried out unconsciously. The second space consists of a network of global workspace neurons that distinguish themselves from the local processors by their reciprocal, long-range anatomical interconnections. Information encoded in the workspace therefore is available to many brain regions at once, including those responsible for motor behavior or verbal report.

Figure 5.5 contains a schematic of the neuronal global workspace.

In this model, sensory stimuli mobilize excitatory neurons with long-range cortico-cortical axons, leading to the genesis of a global activity pattern among workspace neurons. Any such global pattern can inhibit alternative activity patterns among workspace neurons, thus preventing the conscious processing of alternative stimuli.

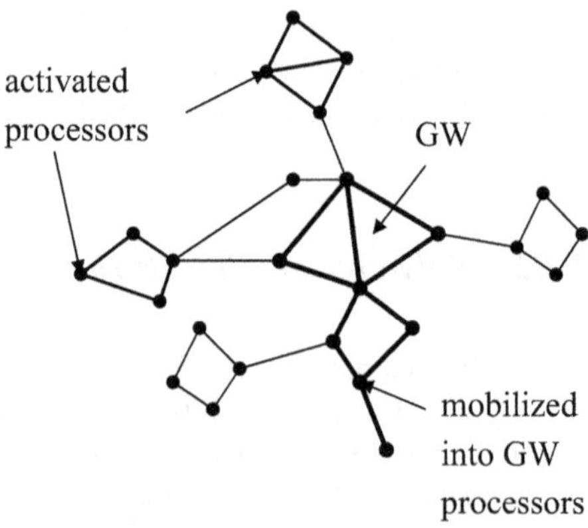

Fig. 5.5 Diagram for global neuronal workspace

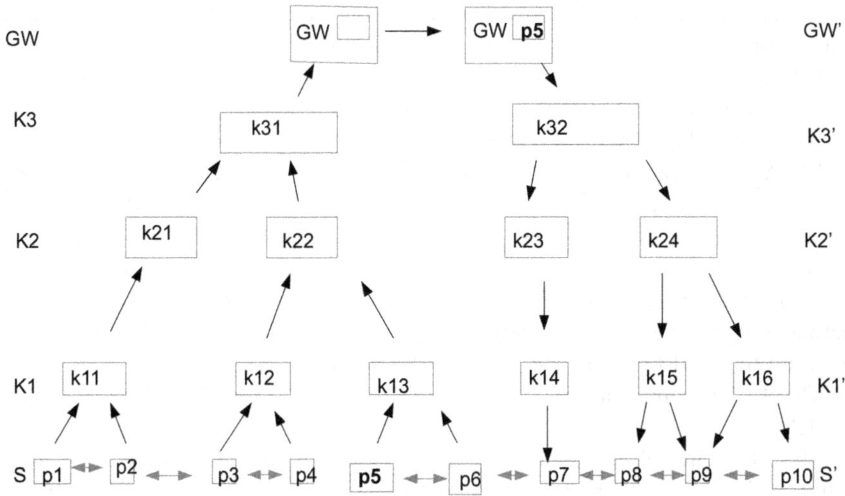

Fig. 5.6 Global workspace architecture

The global neuronal workspace model predicts that conscious presence is a nonlinear function of stimulus significance, that is, a gradual increase in stimulus visibility should be accompanied by a sudden transition of the neuronal workspace into a corresponding activity pattern (Dehaene et al. 2003).

The complementary role of the conscious and unconscious for cognition and self-evolvability or smartness was emphasized. The sway between conscious and unconscious is an important tool for designing creative systems that can autonomously find solutions to highly complex and ill-defined construction problems.

Figure 5.6 illustrates the global work space architecture activity.

When a module for instance p1 invades the workspace, the others, as p2 are blocked at a similar depth.

The rates kij characterizes interactions between conditioning levels Ki and Kj.

In GW theory the processes, p1, p2 and so on, said to be unconscious, compete to enter the global workspace GW. This competition is at several levels.

Such processes are often thought of as memory activities, as for instance episodic or working memories.

Suppose that there are three levels of competition indexed here by K1, K2 and K3 and the competition is won by one process, for instance p5.

Having entered the GW, the wining process becomes the conscious state of the system. This is continuously broadcasted back to the originating processes that change their state according to the conscious state. This results in a new conscious state and so on linking sensory input to memory and conscious states.

LIDA Learning Intelligent Distribution Agent, will be presented as a significant step towards polytopic cognitive architectures.

LIDA, is a conceptual and computational framework for intelligent, autonomous, and conscious software agent that implements some ideas of the global workspace, GW, theory. It appears as an attempt to adopt strategies observed in nature for creating information processing machinery.

LIDA makes use of a partly symbolic and partly connectionist memory organization, with all symbols being grounded in the physical world (Franklin 2006; Baars and Franklin 2007, 2009).

LIDA has distinct modules for perception, working memory, semantic memory, episodic memory, action selection, expectation and automatization (learning procedural tasks from experience), constraint satisfaction, deliberation, negotiation, problem solving, met cognition, and conscious-like behavior. Most operations are done by codelets implementing the unconscious processors, that is, specialized networks of the global workspace theory. A codelet is a small piece of code or program that performs one specialized, simple task.

The LIDA computational architecture is derived from the LIDA cognitive model.

The LIDA model and its ensuing architecture are grounded in the LIDA cognitive cycle. Every autonomous agent, human, animal, or artificial, must frequently sample and sense its environment and select an appropriate response, an action.

Sophisticated agents, such as humans, processes make sense of the input from such sampling in order to facilitate their decision making. The agent's life can be viewed as consisting of a continual sequence of these cognitive cycles. Each cycle constitutes a unit of sensing, attending and acting.

A cognitive cycle can be thought of as a moment of cognition, a cognitive moment.

During each cognitive cycle the LIDA agent first makes sense of its current situation as best as it can by updating its representation of its current situation, both external and internal. By a competitive process, as specified by GW theory, it then decides what portion of the represented situation is most in need of attention. Broadcasting this portion, the current contents of consciousness, enables the agent to chose an appropriate action and execute it, completing the cycle.

Thus, the LIDA cognitive cycle can be subdivided into several phases: perception phase, understanding phase, attention phase that is the consciousness phase, and action selection phase.

Figure 5.6 illustrates some elements of LIDA architecture (Madl et al. 2011; Snaider et al. 2011).

The first module is denoted by S.

The perception is associated to K1. The understanding is associated to K2 and the action selection is associated to K3. The action selection phase of LIDA's cognitive cycle is also a learning phase in which several processes operate in parallel.

The GW space corresponds to the "Self".

The GW mediates between the direct integrative way S $\rightarrow$ K1 $\rightarrow$ K2 $\rightarrow$ K3 and the reverse differentiation way K3′ $\rightarrow$ K2′ $\rightarrow$ K1′ $\rightarrow$ S′ as shown by Figs. 5.6 and 5.7.

Figure 5.7 shows the 4D polytope for LIDA framework.

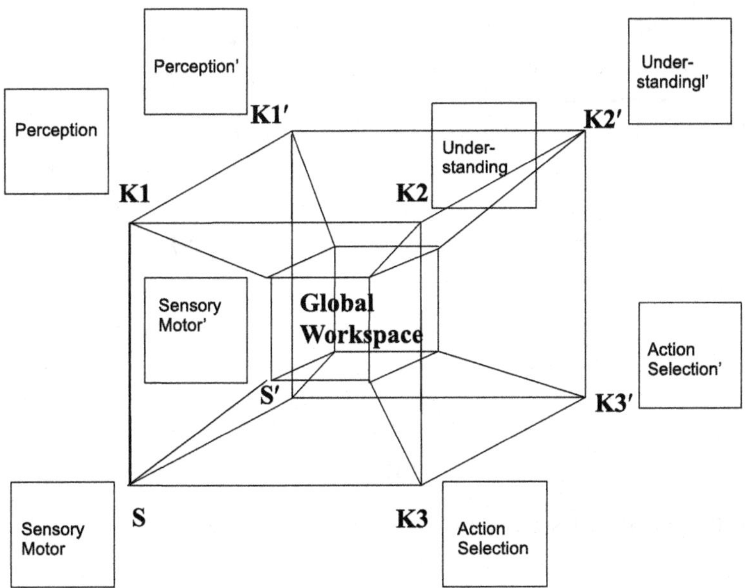

Fig. 5.7 Polytope 4D for LIDA framework

The reverse way allows making use of the developments of the direct way and will offer in a kind of symmetry-breaking or non-commutativity results. The swinging from direct to reverse epistemology is beneficial.

The wave character manifested as swinging behavior is significant for evaluation and creative behavior. The boundaries where creative research grows and new information is created consist of synchronized tendencies. Tendencies to integrate coexist with tendencies to differentiate and it is the blend of both that counts for self-evolvability.

Figure 5.6 suggests a potential application of dual graded graphs as cognitive architecture.

The U operator transfer processes to the GW space while the D operator transfer processes from GW towards field.

Mechanisms used by biology to solve fundamental problems, such as those related to cognitive development, could guide the design of innovative solutions to similar challenges in engineering (Coppens 2004).

Some illustrations are presented in the following.

Figure 5.8 outlines the relation between the polytope 4D and Necker cube.

It helps to clarify the relation between cognitive architectures, high-dimensional polytopes and assimilation vs. accommodation processes (Hemion 2013).

Figure 5.8a shows the Necker cube. Figure 5.8b shows the polytope 4D as a constraint satisfaction network that has the two possible interpretations of the Necker cube as attractor points, where one attractor point is the network state in

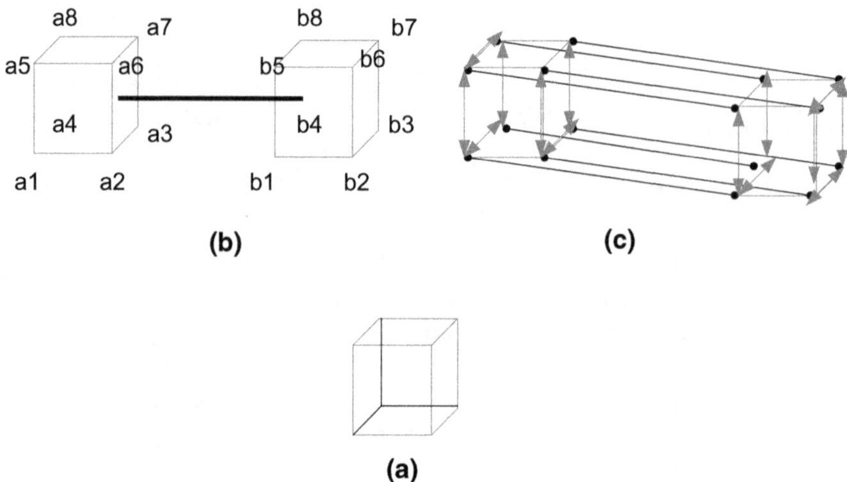

Fig. 5.8 Polytope 4D and Necker cube

which nodes ai, are active and nodes bi, are inactive, and the other vice versa. It is the case distributing-collecting and reverse that is collecting-distributing for process intensification.

Figure 5.8c shows connections between nodes.

Connections shown with arrow markers are excitatory, connections with dot markers are inhibitory. The 8 connections described by the hyper-line have dot markers. Not all connections have been shown. Swing between excitatory and inhibitory supports the new dimension of the cube.

In the interpretation of Rumelhart et al. (1986) the polytopic scheme are like models of the outside world, and processing information with the use of a scheme corresponds to discovering a consistent configuration of schemata which, in concert, offer the best account for the input. This configuration of schemata constitutes the interpretation of the input (Rumelhart et al. 1986). This interpretation of the concept of scheme naturally fits into the language of constraint satisfaction neural networks. The model of the world is composed out of the features represented by the network nodes and constraints for valid configurations are stored in the connection weights. When an input is presented, the network relaxes to an interpretation of the input, or in Piaget's terms, assimilates the input.

The above analysis clarified the significance of duality features for cognitive architectures. Convergence-divergence, reflexive-reflective, unconscious-conscious and other dualities have been pointed out.

The open problem is to impose specific rules of correlation for these duality features.

Since complementary descriptions may appear as irreducible the conceptual integration or coordination of complementary descriptions requires a higher level description, a new dimension that emerges as a new hierarchical level of complexity, as a coordinating center. The more complex is a system the more dimensions and polytopic architectures are necessary to manage complexity.

The swing between complementary ways implies supplementary dimensions to coordinate this swing. Correlating the convergence and divergence processes will allows us to operate control that could not be imposed by only one of these processes in duality.

Damasio elaborated and developed the hypothesis of convergence-divergence zones, CDZ (Damasio 1989, 1994; Meyer and Damasio 2009).

According to this hypothesis, outputs from multiple functional units at one level in the neural structure converge on small assemblies of neurons called CDZ. CDZ detect synchronized firing activities among diverse groups of functional nodes and provide feedback signals returning to these layers that tends to maintain synchronized firing patterns from these groups.

The CDZ is formed by specific cells or neurons whose activity correlates with the presence of a specific object. The specific and invariant responses of these cells attest to a high level of signal convergence in their afferent connectivity and, therefore, demonstrate one of the two key properties of a convergent-divergent neural architecture. However, activating these cells does not complete the task; in other words, it would enable us neither to recognize an object of interest. Reinstating a substantial part of the collection of explicit maps which, in their entirety, represent the meaning of an object of interest, calls for the second key attribute of the CDZ architecture, namely, divergence back projections that establish, in early sensorimotor cortices, the kind of activity patterns for which there is ample experimental evidence. Meyer and Damasio (2009) suggest that special cells or neurons, CDZs, enable the time-locked multiregional retro-activation of explicit maps in early sensorimotor cortices.

Neither direction, divergent or convergent, has much explanatory and predictability value without the other. Every direction alone can account for a limited level of self-organization. The correlation of divergence and convergence directions requires a supplementary dimension that is polytopic architecture.

A 5D polytope inspired by the convergence-divergence model, CDZ is shown in Fig. 5.9.

The main modules are:

- S-Somatosensory
- V-Visual
- A-Auditory
- M-Motor.

Figure 5.9 shows a schematic drawing of cortical organization according to the convergence-divergence model.

The thick hyper-lines denote the construction areas while the thin lines denote the deconstruction or modification areas.

Information is directed from and to the senses and muscles via the primary cortices, the outermost hierarchical level. From there it is first directed to

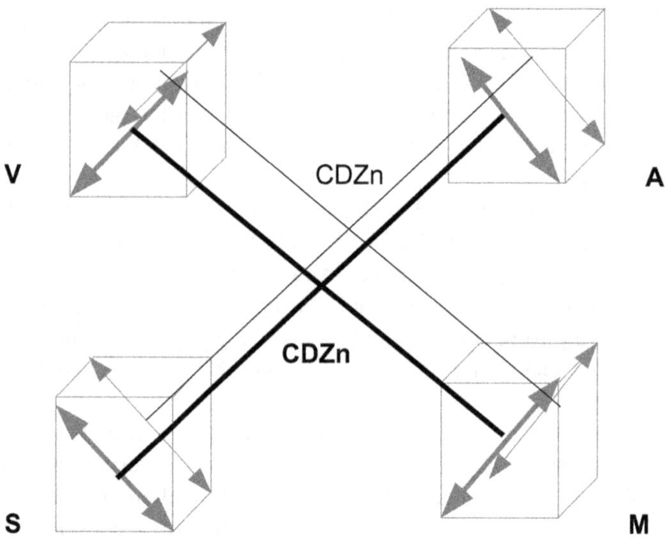

Fig. 5.9 Polytope 5D for convergence-divergence zones model

modality-specific, CDZs and from there onwards to cross-modal CDZs (here depicted generically as CDZn). Information flows both up the hierarchy, for example during perception and learning, as well as down the hierarchy towards the primary cortices, for example during recall (Meyer and Damasio 2009).

References

Anderson, J.R., Lebiere, C.J.: The Atomic Components of Thought. Psychology Press (2014)

Austin, J.: Distributed associative memories for high speed symbolic reasoning. Int. J. Fuzzy Sets Syst. **82**, 223–233 (1995)

Baars, B.J.: A Cognitive Theory of Consciousness. Cambridge University Press, Cambridge (1988)

Baars, B.J.: The conscious access hypothesis: origins and recent evidence. Trends Cogn. Sci. **6**, 47–52 (2002)

Baars, B.J., Franklin, S.: An architectural model of conscious and unconscious brain functions: global workspace theory and IDA. Neural Netw. **20**, 955–961 (2007)

Baars, B.J., Franklin, S.: Consciousness is computational: the LIDA model of global workspace theory. Int. J. Mach. Conscious. **1**, 23–32 (2009)

Berg, C., Saliola F., Serrano, L.: The down operator and expansions of near rectangular k-Schur functions. arxiv e-prints, 1112.4460 (2011)

Björner, A., Stanley, R.P.: An analogue of young's lattice for compositions. arXiv:math.CO/0508043 (2005)

Bringsjord, S., Taylor, J., Wojtowicz, R., Arkoudas, K., van Heuvlen, B.: Piagetian Roboethics via category theory: moving beyond mere formal operations to engineer robots whose decisions are guaranteed to be ethically correct. In: Anderson, M., Anderson, S. (eds.) Machine Ethics. Cambridge University Press, Cambridge (2010)

Carruthers, P.: The Architecture of Mind. Oxford University Press, New York (2006)

Coppens, M.O.: Nature inspired chemical engineering learning from the fractal geometry of nature in sustainable chemical engineering. Fractal Geom. Appl.: A Jubil. Benoit Mandelbrot **72**, 507–532 (2004)

Damasio, A.R.: Time-locked multiregional retroactivation: a systems-level proposal for the neural substrates of recall and recognition. Cognition **33**, 25–62 (1989)

Damasio, A.R.: Descartes' Error: Emotion, Rationality and the Human Brain, 352. Putnam, New York (1994)

Dehaene, S., Sergent, C., Changeux, J.P.: A neuronal network model linking subjective reports and objective physiological data during conscious perception. Proc. Natl. Acad. Sci. U S A **100**, 8520–8525 (2003)

Doumas, L.A.A., Hummel, J.E., Sandhofer, C.M.: A theory of the discovery and predication of relational concepts. Psychol. Rev. **115**, 1–43 (2008)

Epstein, S.: Integration of the cognitive and psychodynamic unconscious. Am. Psychol. **49**, 709–724 (1994)

Epstein, S., Pacini, R., Denes-Raj, V.: Individual differences in intuitive-experiential and analytical- thinking styles. J. Pers. Soc. Psychol. **71**(2), 390–405 (1996)

Evans, J.St.B.T, Over, D.E.: Rationality and Reasoning. Psychology Press, Hove, UK (1996)

Evans, JStBT: Dual-processing accounts of reasoning, judgment and social cognition. Annu. Rev. Psychol. **59**, 255–278 (2008)

Fischer, K.W.: A theory of cognitive development: the control and construction of hierarchies of skills. Psychol. Rev. **87**, 477–531 (1980)

Fischer, K.W., Bidell, T.R.: Dynamic development of action and thought. In: Damon, W., Lerner, R.M. (eds.) Handbook of Child Psychology: Vol. 1. Theoretical Models of Human Development, 6th edn, pp. 313–399. Wiley, New York (2006)

Fischer, K.W., Rose, S.P.: Dynamic growth cycles of brain and cognitive development. In: Thatcher, R.W., Lyon, G.R., Rumsey, J., Kresnegor, N. (eds.) Developmental Neuroimaging: Mapping the Development of Brain and Behavior, pp. 263–283. Academic Press, San Diego (1996)

Fischer, K.W., Stewart, J., Stein, Z.: Process and skill: analysing dynamic structure of growth. In: Riffert, F., Sander, H-J. (eds.) Researching with Whitehead: System and Adventure, pp. 327–367. Verlag Karl Alber, Munich (2008)

Fischer, K.W.: Dynamic cycles of cognitive and brain development: measuring growth in mind, brain, and education. The Educ. Brain: Essays Neuroeduc., 127–150 (2008)

Fomin, S.: Dual graphs and Schensted correspondences. In: Leroux, P., Reutenauer, C., (eds.) Series formelles et combinatoire algebrique., pp. 221–236. LACIM, UQAM, Montreal (1992)

Fomin, S.: Duality of graded graphs. J. Algebraic Combin. **3**, 357–404 (1994)

Frankish, K.: Dual-process and dual-system theories of reasoning. Philos. Compass **5**(10), 914–926 (2010)

Franklin, S.: The LIDA architecture: adding new modes of learning to an intelligent, autonomous, software agent. In: Proceedings of the International Conference on Integrated Design and Process Technology, Society for Design and Process Science, San Diego, CA (2006)

Freeman, W.J.: Neurodynamics: An Exploration of Mesoscopic Brain Dynamics. London, UK, Spinger (2000)

Giraudo, S.: Algebraic and combinatorial structures on Baxter permutations. In: Formal Power Series and Algebraic Combinatorics, FPSAC, **23**, 387–398 (2011)

Grossberg, S.: How does the cerebral cortex work? Learning, attention and grouping by the laminar circuits of visual cortex. Spat. Vis. **12**, 163–186 (1999)

Grossberg, S.: The complementary brain: a unifying view of brain specialization and modularity. Trends Cogn. Sci. **4**, 233–246 (2000)

Grossberg, S.: Adaptive resonance theory: how a brain learns to consciously attend, learn, and recognize a changing world. Neural Netw. **37**, 1–47 (2012)

Halford, G.S., Andrews, G.: Information-processing models of cognitive development. The Wiley-Blackwell Handb. Child. Cogn. Dev. **16**, 697–722 (2010)

Halford, G.S., Wilson, W.H., Phillips, S.: Processing capacity defined by relational complexity. Implications for comparative, developmental and cognitive psychology. Behav. Brain Sci. **21** (6), 803–831 (1998)

Hammond, K.R.: Human Judgment and Social Policy. Oxford University Press, New York (1996)

Hawkins, J., George, D.: Hierarchical Temporal Memory—Concepts, Theory, and Terminology. Numenta, Inc. (2009)

Hemion, N.: Building Blocks for Cognitive Robots: Embodied Simulation and Schemata in a Cognitive Architecture (2013)

Hoffman, M.E.: Combinatorics of rooted trees and Hopf algebras. Trans. Amer. Math. Soc. **355**, 3795–3811 (2003)

Hoffman, M.E.: Rooted trees and symmetric functions: Zhao's homomorphism and the commutative hexagon. arXiv:0812.2419 (2008)

Hummel, J.E., Holyoak, K.J.: Distributed representation of structure. A theory of analogical access and mapping. Psychol. Rev. **104**, 427–466 (1997)

Iordache, O.: Self-evolvable Systems. Machine Learning in Social Media. Springer, Berlin, Heidelberg (2012)

Kegan, R.G.: The evolving self: a process conception for ego psychology. The Couns. Psychol. **8** (2), 5–34 (1979)

Kustrin, D., Austin, J.: Spiking correlation matrix memory. In: International Workshop on Emergent Neural Computational Architectures Based on Neuroscience (1999)

Langley, P., Choi, D.: Learning recursive control programs from problem solving. J. Mach. Learn. Res. **7**, 493–518 (2006)

Lieberman, M.D., Gaunt, R., Gilbert, D.T., Trope, Y.: Reflection and reflexion: a social cognitive neuroscience approach to attributional inference. Adv. Exp. Soc. Psychol. **34**, 199–249 (2002)

Loevinger, J.: Ego Development: Conceptions and theories. Jossey-Bass, San Francisco (1976)

Lücke, J., Keck, C., von der Malsburg, C.: Rapid convergence to feature layer correspondences. Neural Comput. **20**(10), 2441–2463 (2008)

Madl, T., Baars, B.J., Franklin, S.: The timing of the cognitive cycle. PLoS One **6**(4), e14803 (2011)

McLaughlin, G.H.: Psycho—logic: a possible alternative to Piaget' s formulation. Br. J. Educ. Psychol. **33**, 61–67 (1963)

Meyer, K., Damasio, A.: Convergence and divergence in a neural architecture for recognition and memory. Trends Cogn. Sci. **32**, 376–382 (2009)

Mithen, S.: The Prehistory of the Mind. Thames & Hudson, London (1996)

Nestor, A., Kokinov, B.: Towards active vision in the DUAL cognitive architecture. Int. J. Inf. Theor Appl. **11**, 9–15 (2004)

Overton, W.F., Ricco, R.B.: Dual-systems and the development of reasoning: Competence-procedural systems. Wiley Interdisciplinary Reviews: Cognitive Science (2010)

Piaget, J.: Genetic Epistemology. Columbia University Press, New York (1970)

Piaget, J.: The Construction of Reality in the Child. Ballantine Books, New York (1971)

Piaget, J.: Introduction a l'epistemologie genetique. La pensee Mathematique. Presses Universitaires de France, Paris, France (1973)

Piaget, J., Garcia, R.: Psychogenesis and the History of Science. Columbia University Press, New York (1989)

Plate, T.A.: Holographic reduced representations. IEEE Trans. Neural Netw. **6**, 623–641 (1995)

Pollack, J.B.: Recursive distributed representations. Artif. Intell. **46**(1), 77–105 (1990)

Qing, Y.: Differential posets and dual graded graphs. Dissertation, MIT, Cambridge (2008)

Reber, A.S.: Implicit Learning and Tacit Knowledge. Oxford University Press, Oxford, UK (1993)

Rumelhart, D.E., Smolensky, P., McClelland, J.L., Hinton, G.E.: Schemata and sequential thought processes in PDP models. In: Parallel Distributed Processing: Explorations in the Microstructure of Cognition. Vol. 2: Psychological and Biological Models, pp. 7–57. MIT Press, Cambridge, MA (1986)

Schneider, W., Shiffrin, R.M.: Controlled and automatic human information processing: I. Detection, search, and attention. Psychol. Rev. **84**(1), 1–66 (1977)

Selman, R.: The growth of interpersonal understanding: developmental and clinical analyses. Academic Press, London (1980)
Shanahan, M.: A cognitive architecture that combines internal simulation with a global workspace. Conscious. Cogn. **15**, 433–449 (2006)
Shanahan, M.: A spiking neuron model of cortical broadcast and competition. Conscious. Cogn. **17**(1), 288–303 (2008)
Shastri, L., Ajjanagadde, V.: From simple associations to systematic reasoning: a connectionist encoding of rules, variables and dynamic bindings using temporal synchrony. Behav. Brain Sci. **16**(3), 417–493 (1993)
Sloman, A.: Varieties of affect and the CogAff architecture schema. In: Proceedings of the AISB'01 Symposium on Emotion, Cognition, and Affective Computing, York, UK (2001)
Sloman, S.A.: The empirical case for two systems of reasoning. Psychol. Bull. **119**, 3–22 (1996)
Sloss, C.A.: Enumeration of walks on generalized differential posets. M.S. Thesis, University of Waterloo, Canada (2005)
Snaider, J., McCall, R., Franklin, S.: The LIDA framework as a general tool for AGI. In: The Fourth Conference on Artificial General Intelligence, Mountain View, CA (2011)
Sougné, J.: INFERNET: A neurocomputational model of binding and inference. Doctoral Dissertation, University of Liège. Collection PAI, **7** (1999)
Stanley, R.P.: Differential posets. J. Amer. Math. Soc. **1**, 919–961 (1988)
Stanovich, K.E.: Who is Rational? Studies of Individual Differences in Reasoning. Lawrence Erlbaum, Mahwah, NJ (1999)
Stanovich, K.E.: Rationality and the Reflective Mind. Oxford University Press (2010)
Sun, R.: The importance of cognitive architectures: an analysis based on CLARION. J. Exp. Theor. Artif. Intell. **19**, 159–193 (2007)
Sun, R., Merrill, E., Peterson, T.: From implicit skills to explicit knowledge: a bottom-up model of skill learning. Cogn. Sci. **25**, 203–244 (2001)
Sun, R., Coward, L.A., Zenzen, M.J.: On levels of cognitive modeling. Philos. Psychol. **18**, 613–637 (2005)
Wilson, T.D.: Strangers to Ourselves. Cambridge, MA, Belknap (2002)
von der Malsburg, C.: The correlation theory of brain function. In: Domany, E, van Hemmen, J.L., Schulten, K. (eds.) Models of Neural Networks II: Temporal Aspects of Coding and Information Processing in Biological Systems, pp. 95–119. Springer, New York (1994)
von der Malsburg, C.: Dynamic link architecture. In: Arbib, M.A. (ed.) Handbook of Brain Theory and Neural Networks, pp. 365–368. The MIT Press, Cambridge, MA (2002)

Chapter 6
Testing and Designing

6.1 Polytopic Designing

6.1.1 Learning Cycle

Kolb (1984) has described the learning process as a four-phase cycle.

The learner (1st phase) does something concrete or has a specific experience that provides a basis for (2nd phase) the learner's observation and reflection on the experience and his own response to it.

These observations are then (3rd phase) assimilated into a conceptual framework or related to other concepts in the learner's past experience and knowledge from which implications for action can be derived; and (4th phase) tested and applied in different situations.

The four phase cycle is detailed in the following.

1. Experiencing or immersing oneself in the doing of a task is the first stage in which the individual, team, or organization simply carries out the task assigned.
2. Reflection involves stepping back from task involvement and reviewing what has been done and experienced. The skills of attending, noticing differences, and applying terms help identify events and communicate them clearly to others. The learner's paradigm, values, attitudes, beliefs, influences whether he can differentiate certain events. Vocabulary allows communicating perceptions.
3. Conceptualization involves interpreting the events that have been noticed and understanding the relationships among them. At this stage, theory may be particularly helpful as a template for framing and explaining events. Paradigms influence the interpretative range a learner is able to entertain.
4. Planning enables taking the new understanding and translating it into predictions about what is likely to happen next or what actions should be taken to refine the way the task is handled.

O. Iordache, *Advanced Polytopic Projects*, Lecture Notes in Intelligent Transportation and Infrastructure,
https://doi.org/10.1007/978-3-030-01243-4_6

The logic of the learning cycle is to make many small and incremental improvements. When many people make these improvements, major improvements result over time. Likewise, when the learning cycle becomes habitual, the result is continual improvement.

Figure 6.1 illustrates the learning cycle. It is a 2D representation.

The four phases may be associated to S, K1, K2 and K3 frames.

The "Self" performs evaluations allowing balancing the four-phase learning process.

Implementing polytopic projects for learning may start from the direct sequence S → K1 → K2 → K3 and balance this by the reverse or modified sequence: K3′ → K2′ → K1′ → S′. This reverses the arrows direction in K1, K2 and K3.

Modification, either reversion, of the direct sequence frames during reverse processing may happen.

Conditioning from the direct sequence is replaced by de-conditioning in the reverse sequence. It is during these processes of conditioning, de-conditioning and re-conditioning that the differences allowing experience increasing and smartness are generated.

Figure 6.2 shows the polytope 4D for learning cycle.

We can identify:

- S′-Modified Experiencing
- K1′-Modified Reflection
- K2′-Modified Conceptualize
- K3′-Modified Planning.

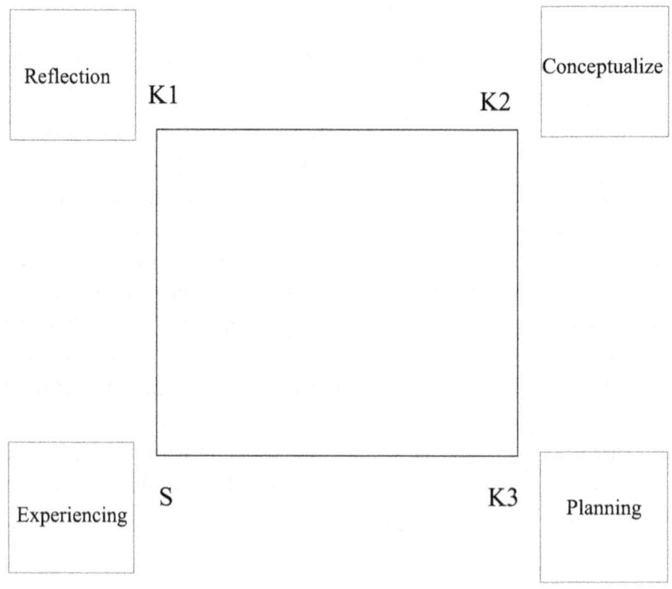

Fig. 6.1 Learning cycle

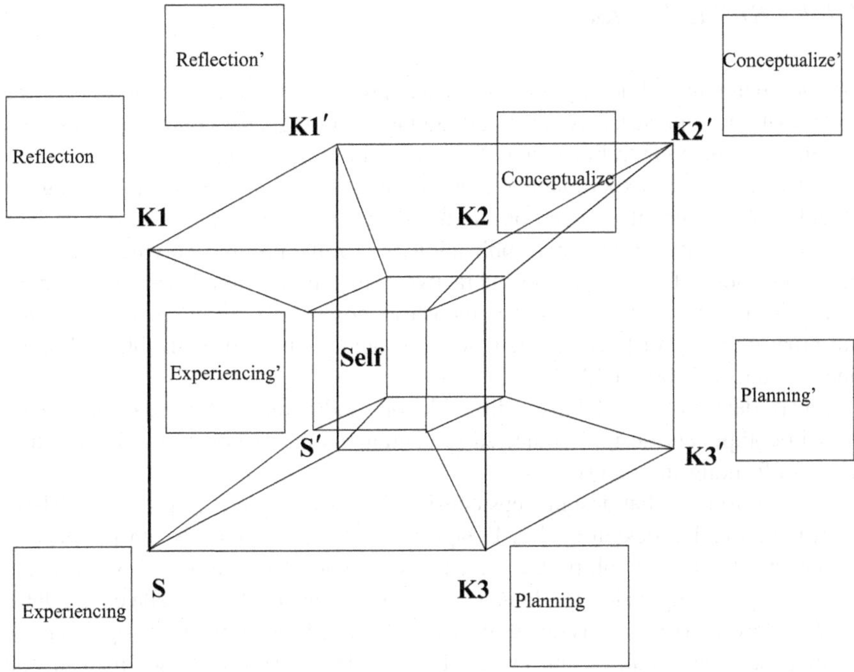

Fig. 6.2 Polytope 4D for learning cycle

Well-known circular design process models, as for example the one of the Institute of Design in Chicago (Research—Analysis—Synthesis—Realization), seem to be variants of Kolb's learning cycles (Jonas 2007).

We can identify:

- S-Research
- K1-Analysis
- K2-Synthesis
- K3-Realization.

This learning cycle in turn, may be related to the very basic cybernetic OODA.

This is an acronym that explains the four steps of decisions making: Observe, Orient, Decide, and Act (Brehmer 2006).

We can identify

- S-Observe
- K1-Orient
- K2-Decide
- K3-Act.

On the basis of such identifications variants of the 4D polytopes for learning cycles may be studied.

6.1.2 RDTE Polytope

Developments in technology and in science have led to a dramatic increase in the complexity of industrial systems. Designing, building and controlling complex systems will be a central challenge for engineers in the coming years.

Smart designs of experiments represent a new approach to problem solving for complexity domain. It is based on the thesis that knowledge can not be a passive reflection of reality, or a passive application of a formal problem solving model, but has to be more of an active and interactive construction. Smart design of experiments is a modern way to cross industrial complexity frontiers by replacing pre-programmed and fixed problem solving methods by self-evolvable and smart ones (Iordache 2009, 2012, 2013).

Smart designs should have adjustable or modifiable architecture since they contain component designs subjected to continuous interaction and reorganization after confronting the reality.

It was observed that design steps consist of typical elementary processes. These form the so called design cycles. Design cycle solves a small design problem or divides it into smaller sub-problems. This observation led to a model which is the repetition of design cycles at different scales (Yoshikawa 1981; Takeda et al. 1990).

Yoshikawa (1981) considered that the so-called ontogenetic design can be decomposed into small design cycles. Each cycle has sub-processes focusing on research, development, testing and evaluation aspects.

We will consider that the design cycle has four basic processes or actions:

- R-Research (includes problem identification and suggesting key concepts to solve the problem).
- D-Development (includes developing alternatives from the key concepts by using design knowledge).
- T-Testing (includes evaluation of alternatives, simulations).
- E-Evaluation and adaptation (includes selection of a candidate for adaptation and modification).

The planned choices for elementary design processes such as research-R, development-D, testing-T and evaluations-E correspond to basic levels.

For sub-processes or the elements of R are R0, R1, R2 and R3, the elements of D are D0, D1, D2 and D3 and so on.

Table 6.1 illustrates several conditioning levels or scales. It is a 2D projection. Here the notations are: R-Research, D-Development, T-Testing, and E-Evaluation.

Then R, D, T and E describe the conditions at the basic conditioning level.

Then R0, R1, R2, R3 are the conditions of R corresponding to the next conditioning level. Then R00, R01, R02, R03 are the sub-conditions of R0 and corresponds to the following level or scale. They were represented as elements of a cyclic loop.

Table 6.1 Array of conditions for RDTE

D11		D12		D21		D22	T11		T12		T21		T22
	D1				**D2**			**T1**				**T2**	
D10		D13		D20		D23	T10		T13		T20		T23
			D							**T**			
D01		D02		D31		D32	T01		T02		T31		T32
	D0				**D3**			**T0**				**T3**	
D00		D03		D30		D33	T00		T03		T30		T33
R11		R12		R21		R22	E11		E12		E21		E22
	R1				**R2**			**E1**				**E2**	
R10		R13		R20		R23	E10		E13		E20		E23
			R							**E**			
R01		R02		R31		R32	E01		E02		E31		E32
	R0				**R3**			**E0**				**E3**	
R00		R03		R30		R33	E00		E03		E30		E33

Table 6.1 contains the network off all the possible conditions, that is, the selected factors, to be grouped in the DOE.

One outcome of the complexity is that the designer may not have time to adequately explore all the design alternatives and select the best alternative.

Consequently, the framework will include only some of the conditions and the corresponding states also. It is a need for modularity and parallelism.

Figure 6.3 shows the polytope 5D for RDTE.

Using the thick hyper-lines, the 5D cube is shown as “square of cubes”.

Here R0, R1, R2, R3 are the conditions of R corresponding to the next level or next scale.

Here R0′, R1′, R2′, R3′ are the modified or reversed conditions of R. The final design should include the acts of redesign which are defined as the act of successive improvements or changes made to a previously implemented design.

Similar observations are valid for D, T and E.

Figure 6.4 shows the polytope 6D for RDTE

With the thick hyper-lines, the 6D cube is shown as “cube of cubes”.

The main steps R, D, T, and E are modified or reversed to the main steps R′, D′, T′, and E′.

This modification may be due to a modified strategy. This strategy is not illustrated in Fig. 6.4 but play the central role in transition from 5D to 6D. In fact exactly such centers allow to increases the dimension.

6.1.3 *Pharmaceutical Polytope*

The self-evolvable and smart designs of experiments may find applications for complex problems as the so-called pharmaceutical pipeline.

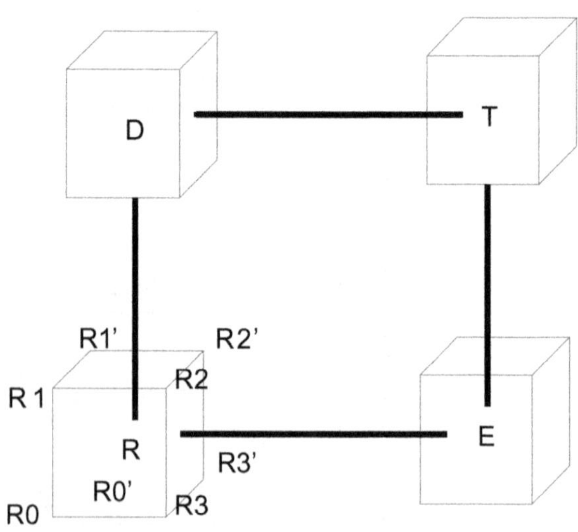

Fig. 6.3 Polytope 5D for RDTE

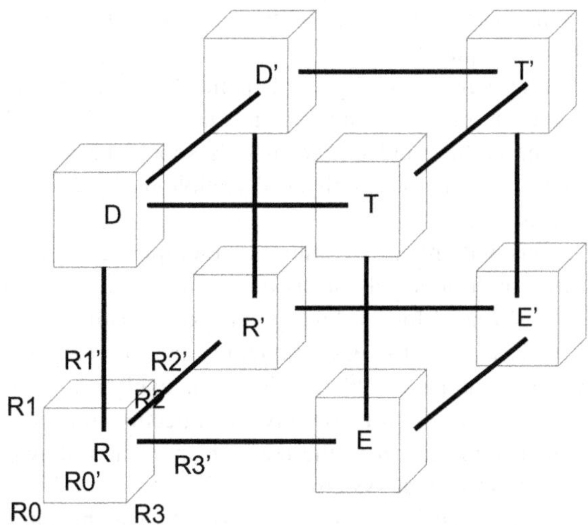

Fig. 6.4 Polytope 6D for RDTE

This refers to the new product, to research and development in pharmaceutical industry.

The typical sequence for new product implementation contains the following main steps:

$$\text{Resources} \rightarrow \text{Discovery} \rightarrow \text{Development} \rightarrow \text{Launching}$$

Biological, chemical and other knowledge resources allow the discovery of drug lead.

The development step includes tests, preclinical, P0, followed by three phases of tests, P1 P2, and P3.

The product launching starts with NDA, New Drug Application, and FDA, Food and Drug Administration, submissions and reviews, and continues with production and marketing steps.

Some areas of pharmaceutical industry are facing a productivity crisis (Woodcock and Woosley 2008). Despite rising investment in pharmaceutical research and development, successful development of new drugs is slowing. The high costs of new drugs development may discourage investment in more innovative, risky approaches in therapeutics.

The FDA, with its dual role of promoting and protecting health is charged with implementing policies that ensures that the benefits of the new products will surpass their risks, while simultaneous promoting innovations that can improve health.

It was observed that chemical and biological systems may have huge behavior spaces and laboratory experiments and models cover only tiny aspects of a system's behavior.

The models often ignore the essential temporal and conceptual space organization of the research and implementation components. Moreover, models and methodologies lack flexibility to adapt and to faster represent more areas of the behavior space.

They neglect synergies—beneficial, nonlinear interactions between systems that cannot be inferred from existing resources and may be missed

The architecture of the models should be in correspondence with that of the studied system within physically, biologically, cognitive or intelligent recognizable spaces.

This will require combining multiple level modeling methods in innovative ways, multiple levels of organization activated both in parallel as in series.

It is a need for modularity and parallelism.

It is a need for new modeling and simulation methods, sufficiently flexible, adaptable and evolvable that is able to explore larger portions of the behavior space, a strong request for cognitive architecture reflecting the essential temporal and spatial organization of the real substrates and allowing autonomy of the new product development system.

Smart DOE, are promising cognitive architectures proposed as new methodologies for problem solving in pharmacology.

The general framework is based on four modules and their self-integrative closure.

The module K0 corresponds to substrate and resources, K1 to discovery step, K2 to developments and testing and K3 to product implementation and launching.

The first module involves resource mining. Resources are material, biological, of knowledge and of intelligent type.

The second module, K1, is that of discovery and involves in this case drug-like molecules discovery, lead discovery and optimization. It may be a standard DOE matrix.

The third module, K2, is that of drug testing and development. It is a meta-design of experiments and for this reason may be denoted by 2-DOE since it refers to processing DOE.

The fourth module, K3, includes application and approval processes, manufacturing, marketing and monitoring of the product.

Each module may involve several sub-modules.

For instance, the module K2 includes a natural four steps cycle P0, P1, P2 and P3.

For the module K3 the NDA step is followed by FDA step this by production and this by product marketing.

Figure 6.5 shows the pharmaceutical pipecycles.

The transition from pipeline to pipecycles proposes a methodology that closes the loop in iterated experimentation in a high-dimensional space.

It is an architectural change from 1D to 2D.

The cycling refers to large cycles for the whole process of four modules or just to one module or sub-module and the corresponding epicycles.

Some cycles may be fully automated if autonomous experimentation methods are used to conduct high-throughput experiments.

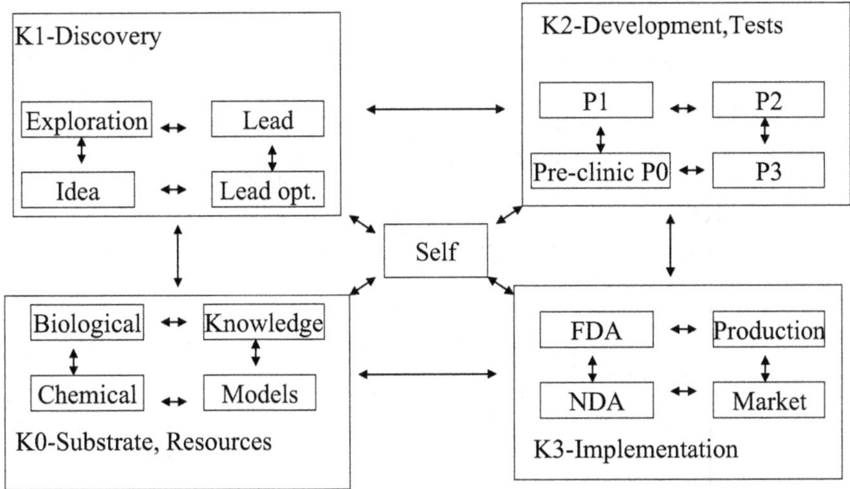

Fig. 6.5 Pharmaceutical pipecycles

Modeling of matrix designs and use of informational criteria accelerate the development of new drugs.

To include the modification that is to evolve, a 3D architecture is necessary.

Self-evolvability and smartness needs 4D architecture.

This will include the "Self".

Figure 6.6 illustrates the pharmaceutical polytope 4D

For multiple scales 5D and 6D architectures are required.

Figure 6.7 illustrates the pharmaceutical polytope 5D.

The notations are:

- Resources-Res
- Discovery-Disc
- Testing-Test
- Implement-Impl.

The main levels are denoted here: Res, Discs, Test and Impl.

Using the thick hyper-lines, the 5D cube is shown as "square of cubes".

Then 0, 1, 2, 3 are the conditions or sub-levels of Res.

Then modified conditions denoted $0''$, $1'$, $2'$, $3'$ are the conditions of Res′.

These kinds of sub-levels are based on the suggestions from dynamic skill theory of development (Fischer 2008; Halford and Andrews 2010).

This decomposition in set-0, mapping-1, system-2 and system of systems-3 has a general validity for complex systems studies.

Similar notations may be introduced for Disc, Test and Impl.

Figure 6.8 illustrates the pharmaceutical polytope 6D.

With the thick hyper-lines, the 6D cube is shown as "cube of cubes".

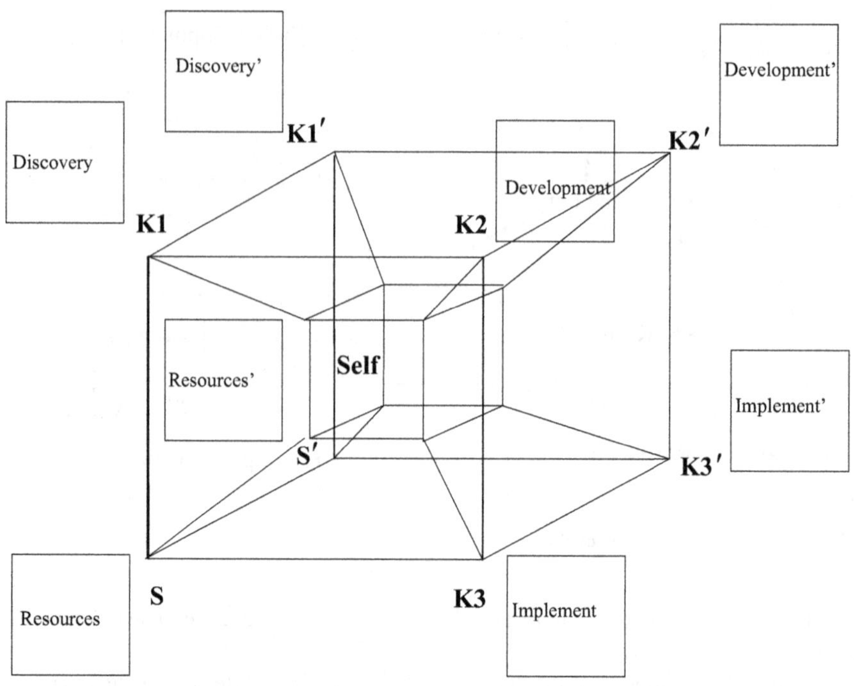

Fig. 6.6 Pharmaceutical polytope 4D

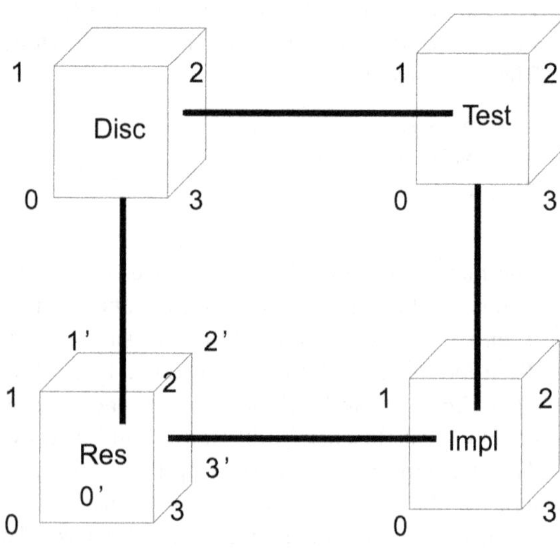

Fig. 6.7 Pharmaceutical polytope 5D

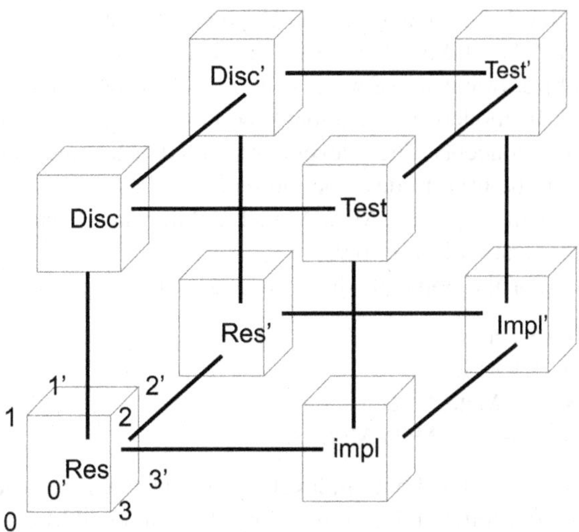

Fig. 6.8 Pharmaceutical polytope 6D

The main levels, Res, Disc, Test, and Impl are modified to the main levels Res′, Disc′, Test′, and Impl′.

6.1.4 Model Driven Architecture

Since the complex systems are structured in levels of reality, associated to multiple scales it is expected that the modeling methods will adopt a similar hierarchical or network architecture.

The four level structures were proposed by the object management group, OMG, to describe informational systems. OMG is an organisation for the standardization in the object oriented field of the software development in which many of the leading software production companies participate (OMG 2008).

For OMG applications in statistics the 0-level refers to data, the 1-level to Models that is models of data, the 2-level to methodologies that is to Meta Models and the 3-level to methodologies that define methodologies that is to Meta Meta Models. Additionally a lower level layer representing physical reality joins the OMG architecture at the 0-level.

The four levels of the polytopic architecture will be denoted by S, K1, K2, and K3 respectively.

K3 is the 3-level, the so-called Meta Meta Models level. One model at level K3 is necessary to define all the K2 level models. The OMG standard for the K3 model called also Meta Object Facility, MOF, is able to define itself (Mellor 2004). This explains why we don't need more levels. MOF is a common framework that is used to define other modeling frameworks within the OMG. K3-model is the language used by MOF to build Meta Models, called also K2-models. Examples of the

2-level, K2-models are the Universal Modeling Language, UML and the relational models. UML has been accepted as a standard notation for modeling object-oriented software systems. Correspondently, at the 1-level, K1, there are UML models and relational models relevant to a specific subject. K1 is based on user concepts. The 0-level denoted K0 or S contains user runtime data or objects. It may be used to describe the real world.

Different Meta Meta Model architectures have been considered as for instance that shown in Fig. 6.9.

For the polytope shown in Fig. 6.9 we can identify:

- S-Data
- K1-Models
- K2-Meta Models
- K3-Meta Meta Models.

The top and bottom levels in architectures are different. In the hierarchical Meta Meta Model architecture every element should be an instance of exactly one element of a model in the immediate next level.

For example, the level K3 could describe elements from the UML Meta Model K2 but not elements from the user models K1. More flexibility is allowed by the centered architectures.

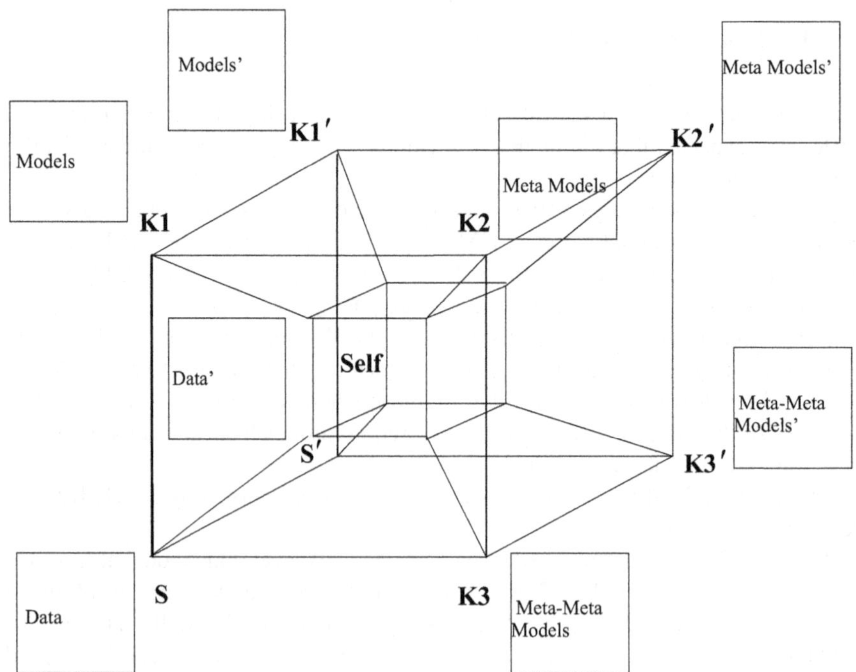

Fig. 6.9 Polytope 4D for Meta Meta Models

Implementing the polytopic project starts from the direct sequence $S \rightarrow K1 \rightarrow K2 \rightarrow K3$ and complements this by the reverse or modified sequence: $K3' \rightarrow K2' \rightarrow K1' \rightarrow S'$.

Adjustment of the direct sequence during reverse processing may arise.

Modeling from the direct sequence is replaced by re-modeling in the reverse sequence.

It is during these direct and reverse processes of modeling, and re-modeling that the variations letting self-evolvability are generated.

Falleri et al. (2007) studied how to use Model Driven Engineering, MDE, in building generic Formal Concept Analysis FCA tools (Ganter and Wille 1999).

Relational Concept Analysis RCA, was introduced as an extension of FCA for taking into account relations between objects. In this way, a concept is described with standard binary attributes but also with relational attributes (Bendaoud et al. 2008; Rouane et al. 2007, 2013).

MDE is a software paradigm introduced to deal more with abstractions rather than code. In a MDE based development every produced or used artifact including code is a Model whose structure is defined by a Meta Model. MDE assume the existence of a single Meta Meta Model which allows defining how a Meta Model is structured.

Hinkelman et al. (2007) studied the relation between engineering design and Meta Model architectures (Mellor 2004; OMG 2008).

The reality corresponds to S. The Model corresponds to K1.

The Model is a simplified representation of reality.

A Model is created with a modeling language. The modeling language specifies the building blocks or elements from which a Model can be made.

There can be different types of modeling languages depending on the kind of Model.

The main types of Models are:

- Graphical model
- Textual description
- Mathematical model
- Conceptual model
- Physical model.

An example could be a table of objects and attributes a mathematical model as in Formal Concept Analysis FCA.

It defines for instance rules to combine object types and relations.

A Meta Model defines the building blocks that can be used to make a model.

It defines:

- Object types that can be used to represent a model
- Relations between object types
- Attributes of the object types
- Rule to combine object types and relations.

An example could be the concept lattice from FCA.

The Meta Model is the abstract syntax while the modeling language is the concrete syntax.

A Meta Meta Model defines the language in which a Meta Model can be expressed.

Meta Object Facility, MOF is a common framework that is used to define other modeling frameworks within the OMG. MOF is itself a Meta Meta Model, a specification describing how one may build Meta Models.

Thus for example MOF provides a specification for how to model the fact that an internet service has service endpoints. Table 6.2 summarizes information for 4 layer Meta Model architectures

Figure 6.10 shows the 4D polytopic project associated to Meta Meta Model approach.

We can identify:

- S-Objects
- K1-Product Details
- K2-Class Attributes Operations
- K3-Meta Class Meta Attributes Meta Operations.

Implementing the polytopic project starts from the direct sequence $S \rightarrow K1 \rightarrow K2 \rightarrow K3$ and complements this by the reverse sequence: $K3' \rightarrow K2' \rightarrow K1' \rightarrow S'$.

Modification of the direct sequence during reverse processing may occur.

Modeling from the direct sequence is replaced by re-modeling in the reverse sequence.

It is during these dual processes of modeling, de-modeling, followed by re-modeling that the variation allowing self-evolvability is generated.

Using the thick hyper-lines, the 5D cube is shown as "square of cubes".

Figure 6.11 shows the Polytope 5D for Meta Model Architecture.

Table 6.2 Meta Meta Model approach

Layer	Description	Examples	FCA
Meta Meta Model K3	Foundation for a Meta Model Architecture Language to describe Meta Models	Meta Class Meta Attribute Meta Operation	Relational Concept Lattices
Meta Model K2	An instance of a Meta Meta Model Language to describe models	Class, Attribute Operation	Concept Lattice
Model K1	An instance of a Meta Model Language to describe information objet domain	Product Details	Object Attribute Tables
Objects S	An instance of a model Define specific information domain	Reality Objects	Reality

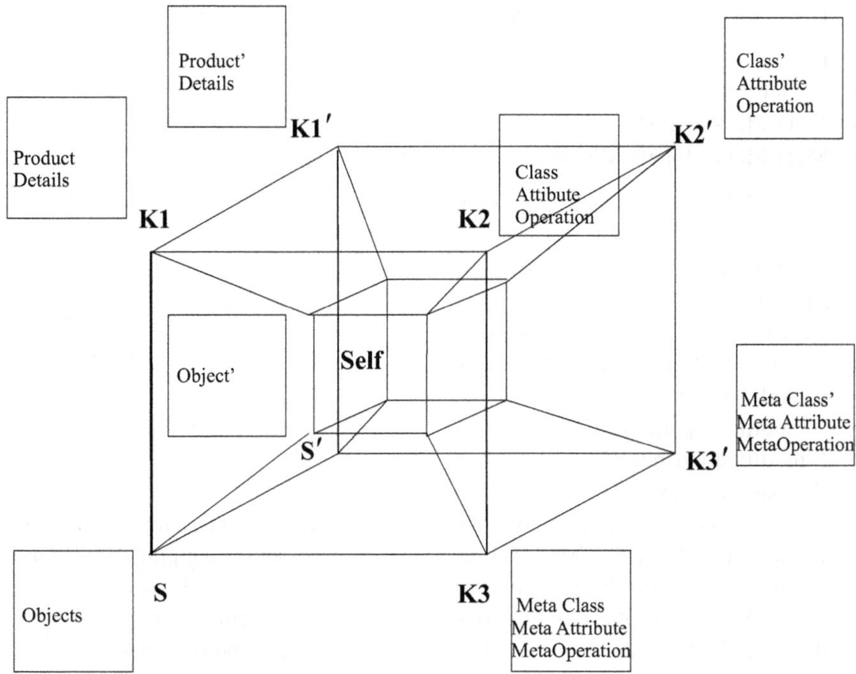

Fig. 6.10 Polytope 4D for Meta Meta Models approach

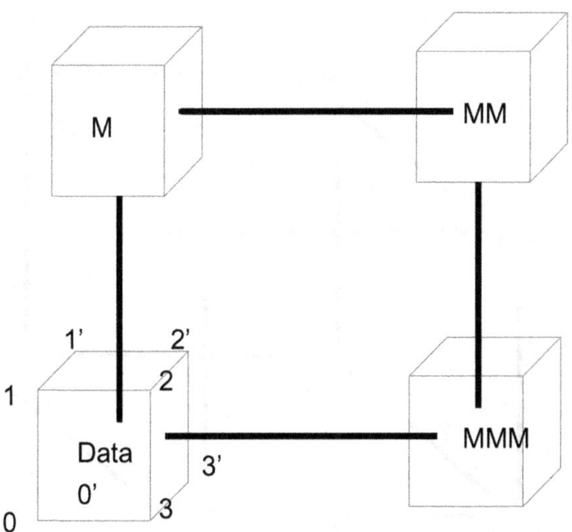

Fig. 6.11 Polytope 5D for Meta Model architecture

The main components of the polytopic architecture are:

- Data
- Models M
- Meta Models MM
- Meta Meta Models MMM.

The sub-levels have been identified as:
0-set
1-operation
2-attribute
3-class
The modified or reversed sub-levels are:
0′-modified set
1′-modified operation
2′-modified attribute
3′-modified class.

These kinds of sub-levels are based on the suggestions from dynamic skill theory of development (Fischer 2008). The decomposition in set-0, mapping-1, system-2 and system of systems-3 has a general validity for different levels of reality.

Figure 6.12 shows the Polytope 6D for Meta Model Architecture

With the thick hyper-lines, the 6D cube is shown as "cube of cubes".

The main steps denoted Data, M, MM, and MMM are modified or reversed to the main steps Data′, M′, MM′, and MMM′.

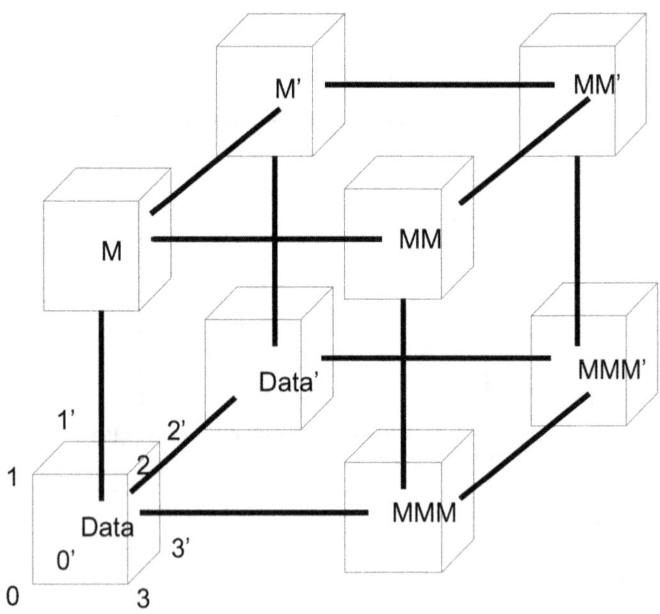

Fig. 6.12 Polytope 6D for Meta Model architecture

6.2 Design Hermeneutics

6.2.1 Hermeneutic Circle

The natural and the human sciences are, different. It makes a difference to the way we view the nature of the design process (Snodgrass and Coyne 1996).

As the activity of designing is a human activity its study belongs within the domain of the human sciences, but also within the domain of the natural sciences. If conceived as reducible to the manipulation of things, it could justifiably be located within the confines of natural science. If conceived as relating to the context of human actions and interactions it must be positioned within the domain of the human sciences.

This is valid also for significant aspects of engineering activity.

Hermeneutic studies attempt to answer the question, how does understanding arise?

The answer is that when we understand, it is because of the working of the hermeneutic circle (Heidegger 1962).

According to Snodgrass and Coyne (1992) the hermeneutic circle is a better method for designing than the dominant method of problem solving because it doesn't destroy the complexity, subtlety, and uniqueness of the design situation, or privilege or preclude aspects of the process, but rather respects their interdependence and interaction.

The hermeneutical circle has to do with the circular relation of the whole and its parts in any event of interpretation. We cannot understand the meaning of a part of an event until we grasp the meaning of the whole; and we cannot understand the meaning of the whole until we grasp the meaning of the parts. That is, we cannot understand the meanings of the stages that make up a sequence until we can locate them in the context of the sequence as a whole; and we cannot understand the meaning of the whole sequence until we understand the meanings of the stages that it comprises. By extension, the meaning of a concept depends on the context, or the horizon, within which it occurs; but this context is made up of the concepts to which it gives meaning.

Any act of understanding involves the interplay of text and context. If you want to build an artifact you need a blue print or a design and you will have a context (Overton 2003).

The whole and the part give meaning to each other; understanding is circular.

Thus we understand what an experiment or a design says to us because of a reciprocal relationship between the whole and the part. These are inseparable in the process of interpretation.

The meaning of the sequence as a whole reflects back and modifies the meanings of its component parts, the stages. The whole can only be understood in terms of its constitutive parts and these parts in turn can only be constructed in terms of the whole which they constitute. This formulation poses a real threat to philosophical concepts that have been considered foundational since Descartes.

The circle involves a logical contradiction: if we must understand the whole before we can understand the parts and yet the parts derive their meaning from the whole, then understanding can never begin. We cannot start with a whole that has no parts; and we cannot start with the parts until we understand the whole. This paradox does not imply that the circle is vicious, but merely that logic is inadequate to the task of understanding the working of understanding. Understanding occurs, so there must be some leap that enables us to understand the whole and the parts at the same time, however contrary to the rules of logic this may seem. Such a leap is a new dimension. An example is the abduction. Abduction is a concept developed by the pragmatist C. S. Pierce to describe a logic where something is interpreted and given new meaning in a new context, as an alternative or complement to inductive and deductive logic which stay in the same context (Overton 2003, 2013).

Looking at this from a different viewpoint, the logical paradox implies that we can only understand the sequence after it has been construed as a whole, so that the meanings of its constituent parts can then be understood in retrospect. Understanding however, does not proceed in this retrospective manner, but at the same time as the event takes place. On a larger scale, we cannot fully understand the parts of a design or an artifact except in the light of the design or artifact as a whole, but we nevertheless understand the parts as we examine them and before we have completed the examination of the whole design.

The person who is trying to understand an experiment, a device or a design is always performing an act of projecting. He projects before himself a meaning for the design as a whole as soon as some initial meaning emerges in the design or experiment. Again, the latter emerges only because he is reading the design with particular expectations in regard to a certain meaning. The working of this fore-project, which is constantly revised in terms of what emerges as he penetrates into the meaning, is the understanding what is there.

When doing a design, or observing an experiment or an artifact, we have initial expectations of what the meaning of the whole will be, and interpret accordingly what we are reading or observing at the moment. We pick up clues and cues from the parts, and from these construct an antecedent formulation of the whole, which then functions in a dialectical fashion to refine and redefine the parts. We move from partial and disjointed insights to an understanding of the whole and back to the yet-to-be-understood portions of the design. As soon as we initially discover some elements that can be understood, we sketch out the meaning of the whole design. We cast forward a preliminary project, which is progressively corrected as the process of understanding advances.

Interpretation brings with it an anticipation, even if informal, of the meaning of the whole; and the light of this anticipation plays back to illuminate the parts. This prior understanding is in turn corrected or confirmed, and gradually specified, as the details react upon it. We project a meaning of the whole even as we begin to read the design or examine an artifact and understand the parts accordingly. This preliminary projection is continually revised as the learner penetrates deeper into the meaning of the parts. The projection, at first unclear and only existing in outline, plays back into the interpretations of the parts, requiring their revision even as the projected meanings itself are continually revised in the light of the interpretation and increasing understanding of the parts. By this process of towards-and-from reflection the understanding of the whole gradually emerges.

Describing what he calls the fore-structure of understanding, Heidegger (1962) says that in any interpretive event, such as understanding language, a design, or the meaning of an object, before we begin consciously to interpret we have already placed the matter to be interpreted in a certain context, viewed it from a pre-given perspective, and conceived it in a certain way.

The process that Heidegger describes is that every revision of the fore-project is capable of projecting before itself a new project of meaning. Rival projects can emerge side by side until it becomes clearer what the unity of meaning is, and interpretation begins with fore-conceptions that are replaced by more suitable ones.

This constant process of new projection is the movement of understanding and interpretation (Snodgrass and Coyne 1996).

6.2.2 *Hermeneutical Polytope*

Progress in hermeneutics of design followed the line of historical hermeneutics (Gadamer 2004). This refers to the fact that the design is embedded in reality and should evolve. Gadamer's hermeneutics is a reaction against claims of finding an objective true interpretation that is a fixed design. Gadamer argues that this is impossible as both the subject and the object of study are situated in history. It is impossible to step outside of this situatedness, no matter what scientific method is used. This view is inspired by Heidegger's notion that to be in the world is to already be situated in history that both the knower (the subject, the designer) and the known (the object that is to be interpreted, the design) have the mode of being of historicity.

Other line of progress follows Ricoeur critical hermeneutics. Ricoeur (1976, 1977, 1990) refers to the critical reexamination of the results of hermeneutical cycle and to the necessary replacing of this cycle by the hermeneutical spiral.

Ricoeur's philosophy builds on Gadamer's historical hermeneutics, but it introduces a critical distancing dimension to interpretation (Jahnke 2012).

It also outlines how re-description plays a crucial part in achieving new meaning.

This approach may be related to Dorst (1997) describing design. Dorst proposed a dual model of design that combines a "reflective practice paradigm", that to a large extent draws on Gadamer's hermeneutics, and a "rational problem solving paradigm" that is problem solving oriented. Such dual models have been studied in Dual Process Theory (Evans 2008; Evans and Over 1996).

Dorst argues that both paradigms are relevant for understanding design practice, but with different emphasis depending on where they occur in the phases of design activity and across design situations. The two paradigms emphasized by Dorst are equally important.

Drawing on Ricouer's critical hermeneutics an alternative approach is to inscribe the rational paradigm within a hermeneutic or reflective framework (Jahnke 2012).

Ricoeur re-introduces epistemology into hermeneutics and establishes a "*long detour*" to understanding in which both are involved: an ontologically derived interpretation and an epistemologically derived reflection which might even be distanced and critical (Bohorquez 2010). These two are intertwined in a hermeneutic spiral that opens up to the "excess of meaning" of the world, rather than locking meaning to established history and tradition. This post-Cartesian understanding of discourse can be seen as a positive, on-going encounter of diverse interpretations-a struggle in which care has to be taken to actually keep tensions and frictions in place because they are fundamental to the process of understanding.

This understanding supports Ricoeur's assertion that critique is fundamental to the goals of keeping communication open and of enhancing the tension needed to generate new meaning.

Observe that all the above described hermeneutical developments, historical and critical hermeneutics or dual models of design imply new dimensions from 2D hermeneutical cycle towards high-dimensional polytope.

Understanding being related to hermeneutical cycle means that polytopic approach will allow a different understanding.

The basic steps of the hermeneutic cycles are: inscription, anticipation, interpreting, and signification.

Doubling from 2D to 4D in the polytopic frame is possible by considering basic steps modification and the "Self".

Figure 6.13 illustrates the hermeneutical polytope 4D.

The notations are:

- S-Inscription-Ins
- K1-Anticipation-Ant
- K2-Interpreting-Int
- K3-Signification-Sig
- S′-Inscription′-Ins′
- K1′-Anticipation′-Ant′
- K2′-Interpreting′-Int′
- K3′-Signification′-Sig′.

To evolve from 4D to 5D each basic step is considered as resulting from 4 sub-steps and the corresponding modifications.

Using the thick hyper-lines, the 5D cube is shown as "square of cubes".

Figure 6.14 shows the hermeneutical polytope 5D.

The main steps are denoted: Ins, Ant, Int and Sig.

The sub-steps have been identified as:

0-set

1-mapping

2-systems

3-system of systems.

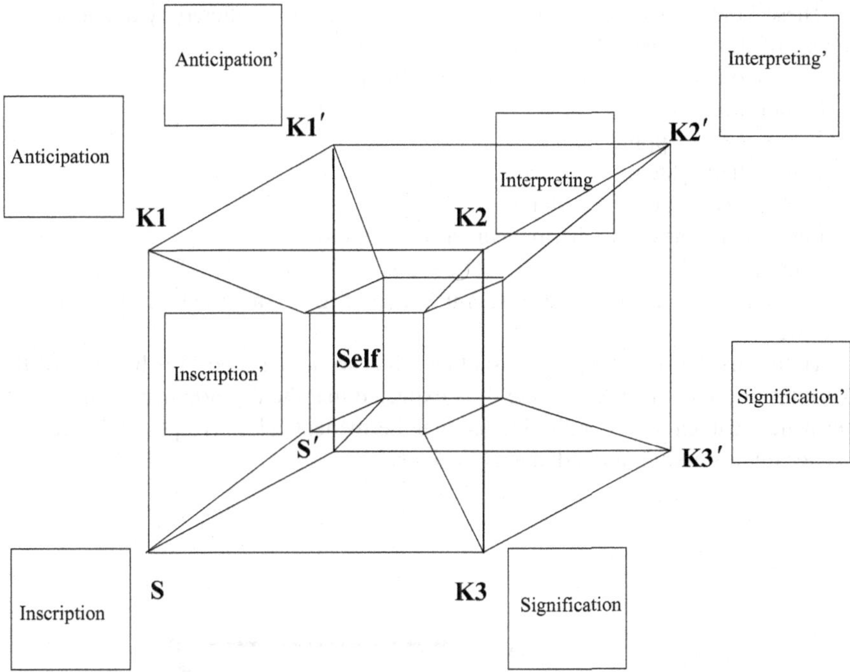

Fig. 6.13 Hermeneutical polytope 4D

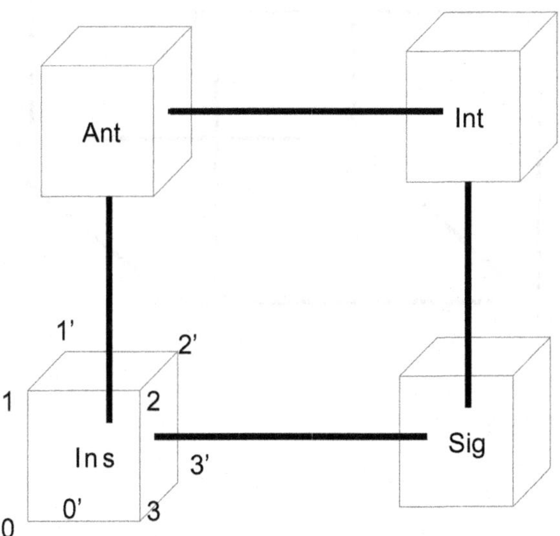

Fig. 6.14 Hermeneutical polytope 5D

These kinds of sub-levels are based on the suggestions offered by dynamic skill theory of development (Fischer 2008).

The modified sub-steps have been identified as:

0′-modified set

1′-modified mapping

2′-modified system

3′-modified system of systems.

Figure 6.15 shows the hermeneutical polytope 6D.

With the thick hyper-lines, the 6D cube is shown as "cube of cubes".

The main steps are denoted: Ins, Ant, Int and Sig are modified to: Ins′, Ant′, Int′ and Sig′.

Traditional hermeneutics suggests that when we understand, it is because of the working of the hermeneutic circle. It is expected that the development from the 2D hermeneutical circle to high-dimensional hermeneutical polytopes will allow a deeper understanding and advanced designs.

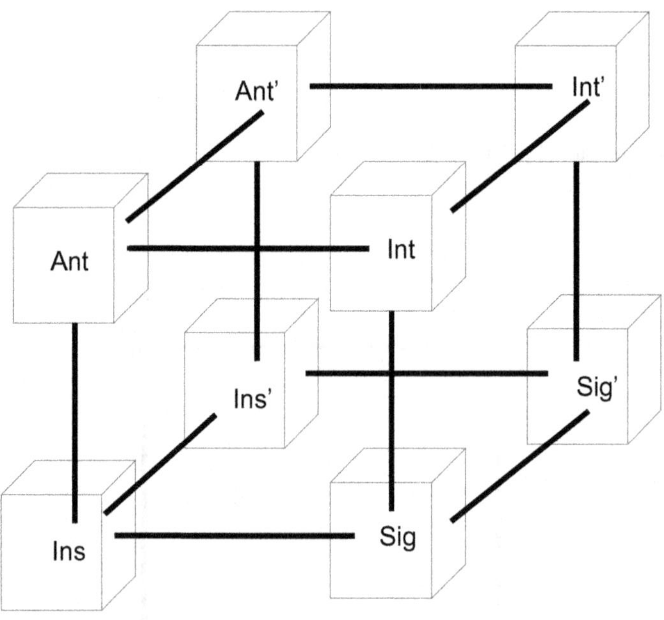

Fig. 6.15 Hermeneutical polytope 6D

References

Bendaoud, R., Napoli, A., Toussaint, Y.: Formal concept analysis: A unified framework for building and refining ontologies. In: International Conference on Knowledge Engineering and Knowledge Management, pp. 156–171. Springer, Berlin, Heidelberg (2008)

Bohorquez, C.: Paul Ricoeur's hermeneutic detours and distanciations: A study of the hermeneutics of Hans-Georg Gadamer and Paul Ricoeur, Ph D thesis, Boston College (2010)

Brehmer, B.: One loop to rule them all. In: Proceedings of the 11th International Command and Control Research and Technology Symposium. Cambridge, UK (2006)

Dorst, K.: Describing Design—a Comparison of Paradigms, Ph D. thesis TUDelft (1997)

Evans, J.S.B.T.: Dual-processing accounts of reasoning, judgment and social cognition. Annu. Rev. Psychol. **59**, 255–278 (2008)

Evans, J.S.B.T., Over, D.E.: Rationality and Reasoning. Psychology Press, Hove, UK (1996)

Falleri, J.R, Arévalo, G, Huchard, M., Nebut, C.: Use of model driven engineering in building generic FCA/RCA tools. In: CLA 2997, vol. 7, pp. 225–236 (2007)

Fischer, K.W.: Dynamic cycles of cognitive and brain development: Measuring growth in mind, brain, and education. The educated brain. Essays in Neuroeducation, pp. 127–150 (2008)

Gadamer, H.G.: Truth and Method, 2nd edn. Sheed and Ward Stagbooks, London (2004)

Ganter, B., Wille, R.: Formal Concept Analysis. Mathematical Foundations. Springer, Berlin (1999)

Halford, G.S., Andrews, G.: Information-Processing Models of Cognitive Development. In: The Wiley-Blackwell Handbook of Childhood Cognitive Development, vol. 16, pp. 697–722 (2010)

Heidegger, M.: Being and Time. Harper and Row, New York (1962)

Hinkelmann, K., Nikles, S., Von Arx L.: An ontology-based modeling tool for knowledge-intensive services. In: Workshop on Semantic Business Process and Product Lifecycle Management (SBPM 2007) 105 (2007)

Iordache, O.: Evolvable Designs of Experiments. Applications for Circuits. Wiley-VCH, Weinheim, Germany (2009)

Iordache, O.: Self-evolvable Systems. Machine Learning in Social Media. Springer, Berlin, Heidelberg (2012)

Iordache, O.: Polytope Projects. Taylor & Francis CRC Press, Boca Raton, FL (2013)

Jahnke, M.: Revisiting design as a hermeneutic practice: An investigation of Paul Ricoeur's critical hermeneutics. Des. Issues **28**(2), 30–40 (2012)

Jonas, W.: Research through DESIGN through research: A cybernetic model of designing design foundations. Kybernetes. **36**(9/10), 1362–1380 (2007)

Kolb, D.A.: Experimental Learning: Experience as the Source of Learning and Development. Prentice Hall, Enlewood Cliffs, NJ (1984)

Mellor, S.J.: MDA Distilled: Principles of Model-driven Architecture. Addison-Wesley Professional, Boston (2004)

OMG: Object Management Group, Software & Systems Process Engineering Meta Model Specification 2.0 (2008)

Overton, W.F.: Understanding, explanation, and reductionism: finding a cure for Cartesian anxiety. In: Reductionism and the Development of Knowledge, pp. 39–62. Psychology Press (2003)

Overton, W.F.: Relationism and relational developmental systems: A paradigm for developmental science in the post-Cartesian era. Adv. Child Dev. Behav. **44**, 21–64 (2013)

Ricoeur, P.: Interpretation Theory: Discourse and the Surplus of Meaning. TCU Press, Fort Worth (1976)

Ricoeur, P.: The Rule of Metaphor: The creation of meaning in language. Routledge, London (1977)

Ricoeur, P.: Time and Narrative, vol. 1. University of Chicago Press, Chicago, IL (1990)

Rouane, M.H., Huchard, M., Napoli, A., Valtchev, P.: A proposal for combining formal concept analysis and description logics for mining relational data. In: International Conference on Formal Concept Analysis, pp. 51–56. Springer, Berlin, Heidelberg (2007)

Rouane, M.H., Huchard, M., Napoli, A., Valtchev, P.: Relational concept analysis: Mining concept lattices from multi-relational data. Ann. Math. Artif. Intell. **67**(1), 81–108 (2013)

Snodgrass, A., Coyne, R.: Models, metaphors and the hermeneutics of designing. Des. Issues **9**(1), 56–74 (1992)

Snodgrass, A., Coyne, R.: Is designing hermeneutical? Archit. Theory Rev. **2**(1), 65–97 (1996)

Takeda, H., Tomiyama, T., Yoshikawa, H., Veerkamp, P.J.: Modeling design processes. Technical Report CS-R9059, Centre for Mathematics and Computer Science (CWI), Amsterdam, Netherlands (1990)

Woodcock, J., Woosley, R.: The FDA critical path initiative and its influence on new drug development. Annu. Rev. Med. **59**, 1–12 (2008)

Yoshikawa, H.: General design theory and a CAD system. In: Man-Machine Communications in CAD/CAM, Proceedings, IFIP W.G5.2, Tokyo, North-Holland, Amsterdam, pp. 35–38 (1981)

Chapter 7
Additive and Subtractive

7.1 Additive Manufacturing

7.1.1 Additive Technologies

Additive manufacturing (AM) techniques are a collection of manufacturing processes which join materials to make physical objects directly using virtual computer data. These processes typically build up parts layer by layer, as opposed to subtractive manufacturing methodologies which create 3D geometry by removing material in a sequential manner. These technologies were also called rapid prototyping, direct digital manufacturing, additive fabrication, additive layer manufacturing, and other different names over the years. An international consensus has coalesced around the use of "additive manufacturing" and "3D printing".

First a 3D CAD file is sliced into a stack of 2D planar layers. These layers are built by the AM machine and stacked one after the other to build up the part (see Fig. 7.1).

Figure 7.1 shows a layered approach to additive manufacturing. Figure 7.1a shows the desired shape and Fig. 7.1b shows the actual shape from additive manufacturing machine.

Obviously the actual shape only approximates the desired shape. More layers will approximate better.

The construction by layers may be completed or replaced by construction by voxels.

A voxel is a volumetric pixel, used to define the fundamental unit of digital space. Voxels can be both digital and physical. Digital voxels are computational representations in 3D models. Physical voxels may be comprised of materials as diverse as basic raw materials nanomaterials, integrated circuits, biological materials, and micro-robotics, among others.

Material jetting is the use of inkjet printers or other similar techniques to deposit droplets of build materials that are selectively dispensed through a nozzle or orifice to build up a 3D-structure. In most cases these droplets are made up of photopolymers or wax-like materials to form parts or investment casting patterns

O. Iordache, *Advanced Polytopic Projects*, Lecture Notes in Intelligent Transportation and Infrastructure,
https://doi.org/10.1007/978-3-030-01243-4_7

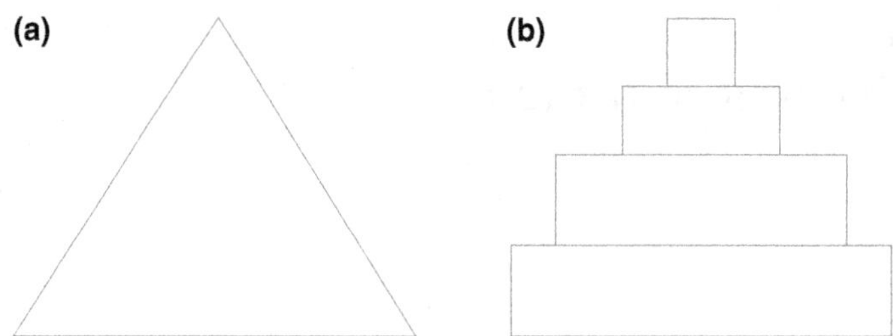

Fig. 7.1 Layered approach to additive manufacturing

respectively. These processes are 3D-printing machines, as they use inkjet and other printing techniques to build up 3D-structures.

Binder jetting techniques also use nozzles to print material, but instead of printing with the build material, the printed material is glue, which holds powder together in the desired shape. A binder jetting process starts by first depositing a thin layer of powder. A print head is then used to print a glue pattern onto the powder, thus forming the first layer. A new layer of powder is deposited and glue is printed again. This pattern is repeated until the part is completed.

The largest installed base of AM techniques is based upon material extrusion (Stucker 2012). Material extrusion machines work by forcing material through a nozzle in a controlled manner to build up a structure. The build material is usually a polymer filament which is extruded through a heated nozzle. After a layer of material is deposited by the nozzle onto a platform, the platform either moves down or the nozzle moves up, and then a new layer of material is deposited.

Sheet lamination techniques work by cutting and stacking sheets of material to form an object. This approach has been used with paper, plastic and metal sheets to build up wood-like, plastic and metal parts respectively. A binder is typically used to bond paper and plastic sheets; whereas welding or bolting of sheets together is typically used for metals. Additionally, sheet lamination has been used with ceramic and metal tapes to build up structures which are later fired in a furnace to achieve a dense part.

Several AM techniques have been modified to work at a small scale to deposit passive electronic structures as conductors, insulators, resistors, antennas, and so on. These techniques are often known as direct write techniques and, make use of electronic inks that contain nano-particles or other additives that result in electronic properties after drying, thermal decomposition or other post-treatment. By combining direct write techniques with other AM techniques it becomes possible to create multi-functional 3D-embedded electronic structures on a layer-by-layer basis that combine structural, thermal, electronic and other functions into a single component (Stucker 2012).

The AM may be described using Young diagrams (Forcey et al. 2007, Iordache 2012).

In this case the AM is performed voxel by voxel.

The AM procedure is described by the product and coproduct of diagrams.

Figure 7.2 shows a Young diagram for planar dendrites growth. Here A and B denote the dendrites.

Here product $\otimes_1$ denotes the horizontal stacking while $\otimes_2$ denotes the vertical stacking.

Figure 7.3 shows a 3D-Young diagram for spatial growth.

In this case $\otimes_1$ denotes the z-axis stacking that is the vertical concatenation.

The product $\otimes_2$ denotes the y-axis stacking that is the horizontal concatenation.

The corresponding matrices are:

$$A = \begin{bmatrix} 4 & 3 & 1 & 1 \\ 4 & 2 & 1 & 1 \\ 3 & 2 & 1 & \\ 1 & 1 & 1 & \end{bmatrix} \quad B = \begin{bmatrix} 3 & 1 \\ 2 & 1 \\ 1 & 1 \end{bmatrix} \tag{7.1}$$

$$A \otimes_1 B = \begin{bmatrix} 4 & 3 & 1 & 1 \\ 4 & 2 & 1 & 1 \\ 3 & 2 & 1 & \\ 3 & 1 & 1 & \\ 2 & 1 & & \\ 1 & 1 & & \\ 1 & 1 & & \end{bmatrix} \quad A \otimes_2 B = \begin{bmatrix} 4 & 3 & 3 & 1 & 1 & 1 \\ 4 & 2 & 2 & 1 & 1 & 1 \\ 3 & 2 & 1 & 1 & 1 & \\ 1 & 1 & 1 & & & \end{bmatrix} \tag{7.2}$$

Only the non-zero entries of the matrices are shown.

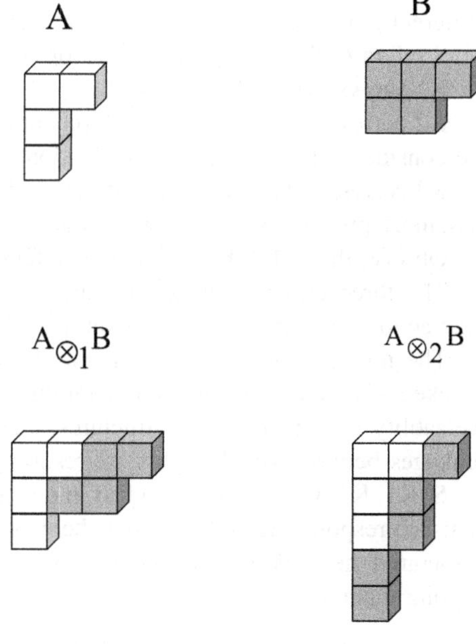

Fig. 7.2 Young diagrams for planar growth

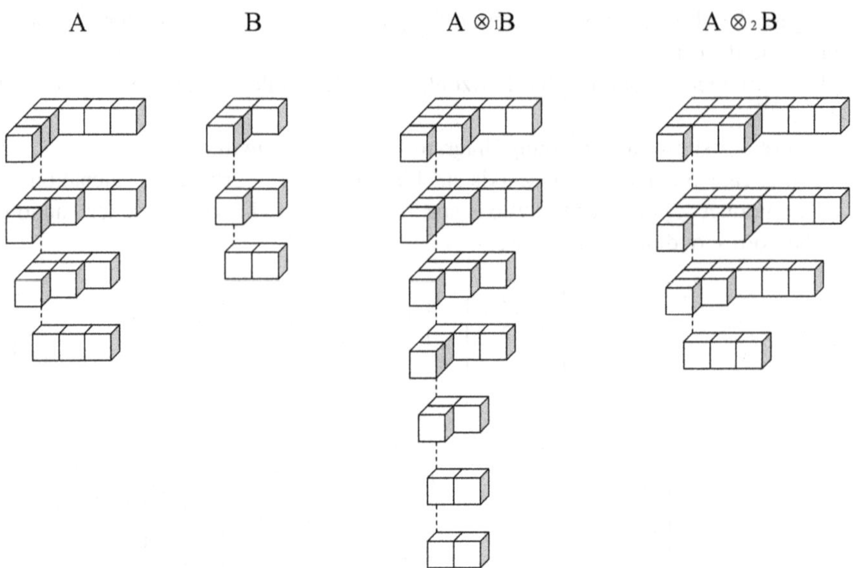

Fig. 7.3 Young diagrams for spatial growth

Such models describe the evolution of certain crystals formed in solutions. The fact that at certain temperatures the usual regular increase in size of the crystal became a rhythmic growth find natural explanation with this type of models (Ferreiro et al. 2002; LaCombe et al 2002).

Figure 7.4 shows a 4D polytope for spatial growth.

Figure 7.4 illustrates the case study of spatial growth and shows the place of different structures in the 4D polytope frame.

S refers to the elementary voxels or cells allowing the dendrites growth.

K1 shows the chains of voxels.

K2 corresponds to planar development of dendrites. Several possibilities should be considered for different crystallization degrees and shapes.

K3 corresponds to assembly of planar dendrites. It is a self-templating and self-assembly process. Self-folding is a process that converts 2D objects in 3D shapes.

Observe that in K1, K2 and K3 may refer to different types of physical interactions.

The three dimensional configuration of one material held together by voxels interactions may be less stable than the other possible candidates.

The multiplicity ensures the flexibility of the K3 level. The "Self" organized stacked structures with weak but multiple linkages between voxels layers allow versatility and smartness of structures. New functional materials due to the weak linkages between voxels layers are resulting.

S, K1, K2 or K3 may be studied in detail as polytopic architecture of sub-levels. This corresponds to the fact that there is not only one voxel at K2 level but a structured network of interacting voxels for instance all structures shown in a Young diagram.

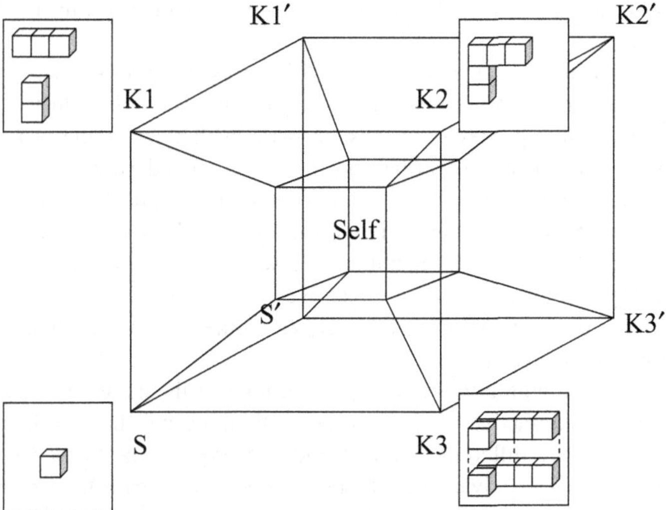

Fig. 7.4 Polytope 4D for spatial growth

Experimental observations suggest the potential to use process parameter change both to better understand the phenomena of voxels growth, the side-branch formation and the selection process in dendrites growth, and as a way to use pressure steps or oscillations to control or engineer dendrites microstructures.

7.1.2 Addition and Subtraction

Subtractive processes include those that take away material to create a workpiece's shape. The traditional metal removal processes such as milling, turning, grinding, electrical discharge machining and so on, fall into this category. Additive processes include those that add material, usually layer by layer, to build up a new desired shape. Direct metal laser sintering, stereolithography and 3D printing are examples.

The focus should be on the complementary nature of these seemingly opposite approaches to form a workpiece.

Additive manufacturing create objects from the bottom-up, whereas subtractive manufacturing are working top-down. Subtractive and additive processes can be combined to develop innovative manufacturing methods that are superior to conventional methods. Using additive and subtractive processes together helps us make things that could not be made as well by other means. One-piece objects with complex, enclosed interior spaces and detailed, external features are examples. This might be a mold in which cooling lines were produced with direct metal laser sintering, but the cavity was finished with high speed milling. Another example could be using an additive process to build up a turbine blade while creating curved air cooling channels within, followed by creep-feed grinding to finish contours on an outer surface.

A high-dimensional heat transfer method is necessary in this case.

It is true that additive processes, which have come recently in industry, are likely to replace subtractive processes at least in part. In that sense, the processes will compete with each other. Nevertheless, the focus should stay on the creativity enabled by their combination. This creativity is both an appeal and a challenge to the imagination. Sometimes the designer will take the lead in exploiting this creative freedom. Other times the manufacturer will have to show the way by suggesting new possibilities to bring inventive part shapes into reality. Dynamic tension between designer and manufacturer is a normal sign of valid innovation and should be maintained (Dorst 1997).

Successfully combining these processes means that the additive or constructive thinking and the subtractive or deconstructive thinking will have to blend. These modes of thought are different. With an additive process, the blank workpiece is empty space. This is an advantage when components are characterized by internal voids or enclosed cavities corresponding to the absence of material. The inherent efficiency lies in the fact that unneeded material is left out from the start. Many parts still will be primarily large pieces of matter. Starting with a solid mass will be the advantage, because material that will yield the finished shape is already present in a known condition.

The cooperation between additive and subtractive steps implies an emerging high- dimensional manufacturing. Correlating the additive and subtractive processes helps us make objects that could not be made by only one of these processes.

Associating additive processes as an option is an opportunity not an impediment for prototypes. Most of them have been moving toward greater diversity in manufacturing capability for a long time now anyway. The key is offering a sum of specialties that add up to greater value for the industry.

Some implications of combinatorial analysis models for addition and subtraction technologies will be presented in what follows.

A generalization of conventional Young lattices has been proposed by Gouyou-Beauchamp and Nadeau (2007). They introduced layered graphs, ribbons and signs.

A ribbon is a connected skew shape that does not contain a 2 by 2 square of cells.

The size of a ribbon is its number of cells.

The height h(r) of a ribbon is its number of rows minus 1.

The sign of a ribbon is $(-1)^{h(r)}$.

Ribbons may be part of additive manufactured objects. The ribbons correspond to deposited layers of different materials.

Figures 7.5 and 7.6 show examples of ribbons.

The ribbon shown in Fig. 7.5 has size 7, height 2 and sign +1.

The ribbon shown in Fig. 7.6 has size 9, height 3 and sign −1.

Instead of layer by layer AM may be performed ribbon by ribbon.

Both methods may be applied and correlated.

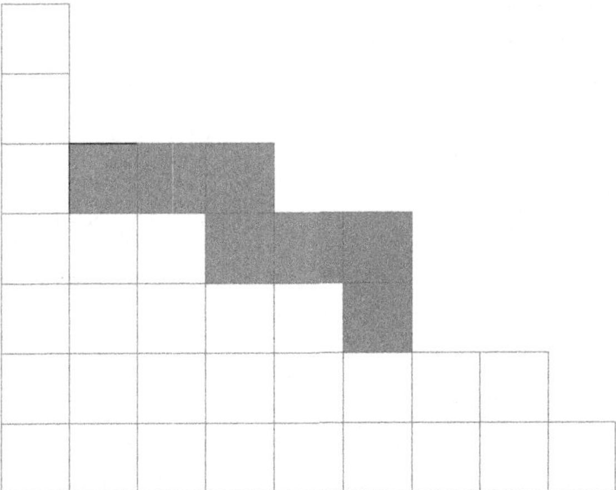

Fig. 7.5 Ribbon example-a

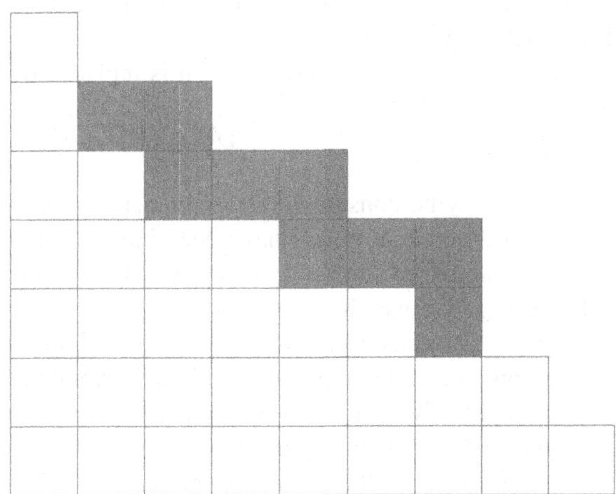

Fig. 7.6 Ribbon example-b

Figure 7.7 shows the first levels of the ribbon graph (Gouyou-Beauchamp and Nadeau 2007):

A layered graph is self-dual if the commutation conditions are:

$$D_iU_i - U_iD_i = r_iI \quad (7.3)$$

$$D_iU_j - U_jD_i = 0 : \text{if } i \neq j \quad (7.4)$$

Here U_i (respectively D_i) denote the restriction of the operator U (respectively D) to the ith level of the graph, I_i denote the identical operator at the same level and r_i are positive integers.

Notice that the conventional approach of Stanley (1988) and Fomin (1994) considered that edges exist only between consecutive layers and the sign function is constant and equals one.

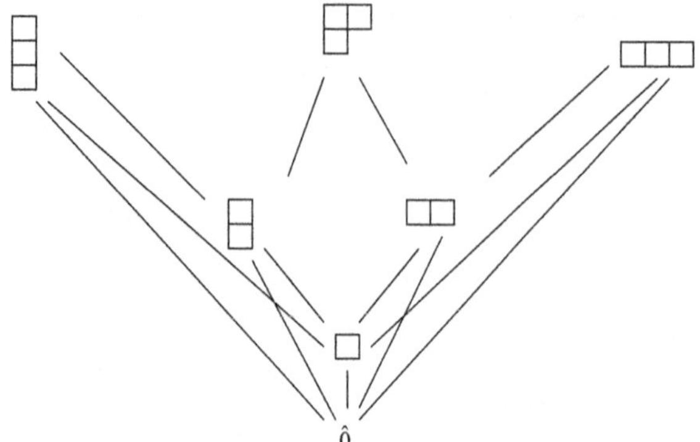

Fig. 7.7 First levels for the ribbon graph

Equations as (7.3) and (7.4) generalize commutation conditions for differential posets.

Another generalization is done by the quantized dual graded graphs studied by Lam (2010) (Appendix A).

For these the commutation relation is replaced by:

$$DU - qUD = rI \tag{7.5}$$

Here q may be considered as a parameter which the graph depends upon.

Signed differential posets have been introduced by Lam (2008).

Lam generalized the dual graded graphs introduced by Fomin (1994) by considering negative weights.

Signed dual graded graphs allows to model additive and subtractive processes

For signed enumerations one deals with weights plus or minus one.

The negative weight may be interpreted as a subtractive step.

Figure 7.8 shows the signed U-graph.

Figure 7.9 shows the signed D-graph.

Figure 7.10 shows the commutation condition for signed graphs.

As observed the commutation relation generalizes that of Fomin (1994).

Layered graphs, ribbons and signed graphs are basic tools for the design and study of additive/subtractive processes.

7.1.3 4D Printing

The 4-dimensional printing, 4D printing, uses the same techniques of 3D printing through computer-programmed deposition of material in successive layers to create a three-dimensional object. However, 4D printing adds the dimension of transformation over time. It is therefore a type of programmable matter (PM), wherein after the fabrication process, the printed product reacts with parameters within the environment (humidity, temperature, pressure and so on) and changes its form

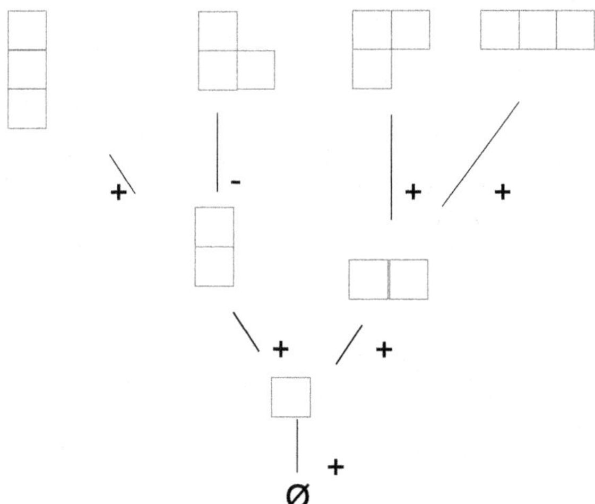

Fig. 7.8 Signed U-graph

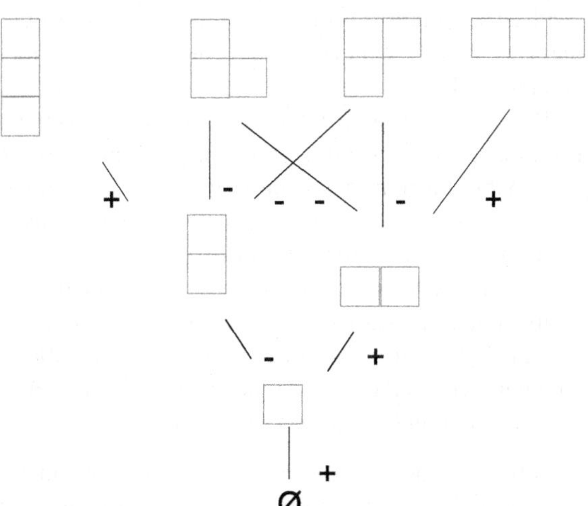

Fig. 7.9 Signed D-graph

accordingly (Campbell et al. 2014). The ability to do so arises from the large number of configurations at a micrometer resolution, creating solids with engineered molecular spatial distributions and thus allowing unprecedented multi-functional performance.

The 4D printing is a new advance in bio-mimetic fabrication technology, rapidly emerging as a new paradigm in disciplines such as bioengineering, materials science, chemical engineering, and computer sciences.

Rather than construct a static material or one that simply changes its shape, the development of adaptive, bio-mimetic composites that reprogram their shape, properties, or functionality on demand, based upon external stimuli, is proposed. By integrating the abilities to print precisely, 3D hierarchically-structured materials, to

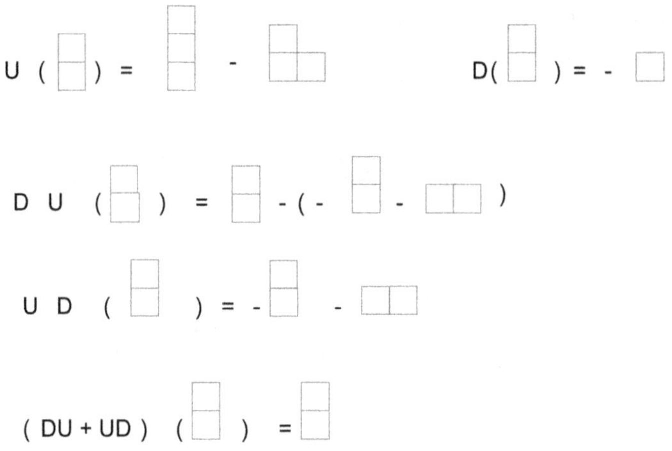

Fig. 7.10 Commutation condition for signed graphs

synthesize stimuli-responsive components and to predict the temporal behavior of the system, it is expect to build the foundation for the new field of 4D printing (Tibbits 2014).

PM is matter which has the ability to change its physical properties as for instance shape, density, modulus, conductivity, optical properties in a programmable fashion, based upon user input or autonomous sensing.

PM is thus linked to the concept of smart material which inherently has the ability to perform information processing.

The ability to create one object that responds to light by changing its color, to temperature by altering its permeability, and to an external force by hardening its structure, becomes possible through the creation of smart materials that are simultaneously adaptive, flexible, lightweight, and strong.

PM may come in at least two forms:

(1) objects made of pre-connected elements that are 4D printed or otherwise assembled as one complete structure for self-transformation
(2) unconnected voxels that can come together or break apart autonomously to form larger programmable structures.

PM encompasses, a range of technological capabilities-including 3D printing, micro-robotics, smart materials, nanotechnology, micro-electromechanical systems (MEMS) and micro-energy-chemical systems (MECS) (Mahalik 2005).

Figure 7.11 illustrates the polytope for 4D printing.

S refers to the elementary voxels or cells necessary for manufacturing.

K1 shows the linear printing. K2 corresponds to planar or layer developments. K3 corresponds to assembly of planar deposits that is to multiple layers. It is a self-templating and self-assembly process.

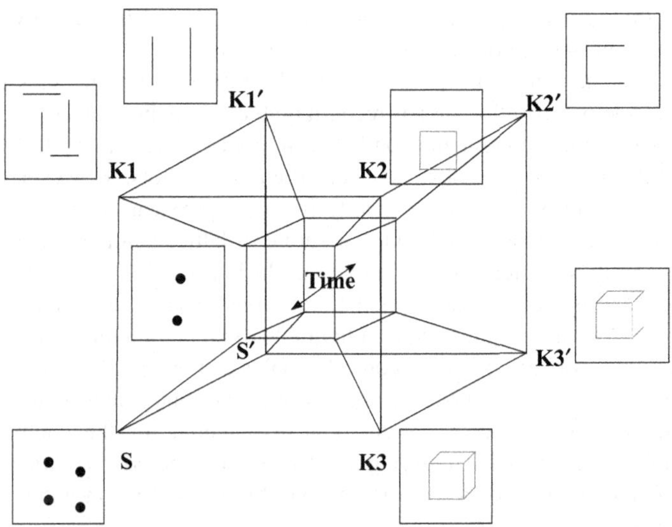

Fig. 7.11 Polytope for 4D printing

The "Self" organizes stacked structures with weak but multiple linkages between layers allow versatility of structures. New functional materials due to the weak linkages between layers are resulting.

7.2 Modular Reconfigurations

7.2.1 Hexagonal Configurations

Features that are desirable for the smart industry would relate to being flexible and reconfigurable, low cost, adaptive or transformable, and agile. One of the ways to achieve such functionalities would be to apply modular structure with respect to product, process technology and organization.

A self-reconfigurable robot is a robot built from potentially many modules which are connected to form the robot. Each module has sensors, actuators, processing power, and means of communicating with connected modules. The robot autonomously changes shape by changing the way these modules are connected. Self-reconfigurable robots have a high degree of robustness, versatility, scale extensibility, and adaptability. This makes these robotic systems interesting to study.

Self-reconfigurable robots are built from robotic modules typically organized in a lattice. The robotic modules themselves are complete, although simple, robots and have onboard batteries, actuators, sensors, processing power, and communication capabilities. The modules can automatically connect to and disconnect from neighbor modules and move around in the lattice of modules. The self-reconfigurable robot as a whole can, through this automatic rearrangement of modules, change its own shape to adapt to the environment or as a response to new tasks.

Potential advantages of self-reconfigurable robots are high versatility and robustness. The organization of self-reconfigurable robots in a lattice structure and the emphasis on local communication between modules mean that lattice automata are a useful basis for control of self-reconfigurable robots. However, there are significant differences which arise mainly from the physical nature of self-reconfigurable robots as opposed to the virtual nature of lattice automata. The problems resulting from these differences are mutual exclusion, handling motion constraints of modules, and unrealistic assumption about global, spatial orientation.

Despite such problems the self-reconfigurable robot community has successfully applied lattice automata to simple control problems. For more complex problems hybrid solutions based on lattice automata and distributed algorithms are used. Hence, lattice automata have shown to have potential for the control of self-reconfigurable robots, but a unifying implementation based on lattice automata solving complex control problem, running on physical self-reconfigurable robot is to be demonstrated.

Some modular robots needs hexagonal configuration. For example, we refer to the ATRON modules (Jorgensen et al. 2004; Christensen and Stoy 2006).

A two module meta-module 2D lattice suggests studying possible hexagonal configuration (Propp 1989).

Three configuration a, b and c are presented in the following.

Figure 7.12 shows the hexagonal configuration-a.

Here the circle denotes a module.

Figure 7.13 shows the hexagonal configuration-b.

Figure 7.14 shows the hexagonal configuration-c.

Figure 7.15 shows the complete hexagonal configuration.

For simplicity the dots • instead of the circles denotes the modules, as ATRON for example.

The hexagonal configurations may be studied as dual graded graphs (Appendix A).

Figure 7.16 shows the U and D graphs for hexagonal configurations.

The weights indicated on the connections between two dots or two modules show the number of ways we can obtain one from another in up U and down D direction (Fomin 1994).

Figure 7.17 illustrates the commutation condition for the graphs shown in Fig. 7.16.

Commutation relations allow specifying trajectories for modular coupling and de-coupling processes.

7.2.2 High-Dimensional Configurations

Reconfigurable systems may be built by cube-shape modular robots (Rus and Vona 2000; Aloupis et al. 2009; Gilpin and Rus 2010).

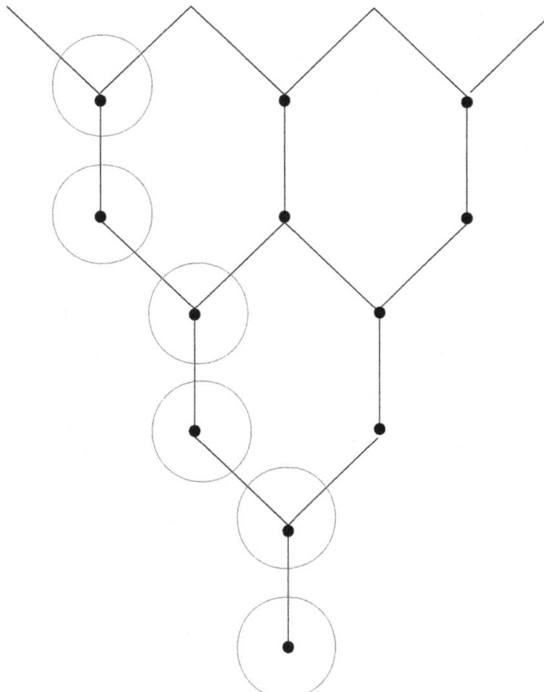

Fig. 7.12 Hexagonal configuration-a

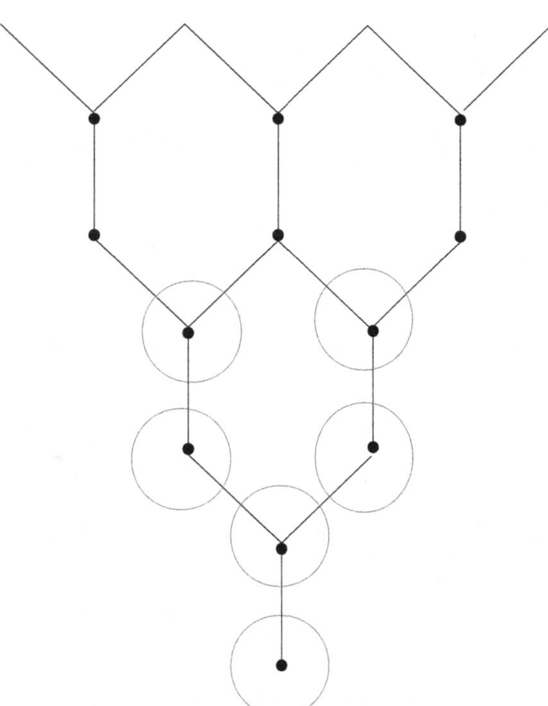

Fig. 7.13 Hexagonal configuration-b

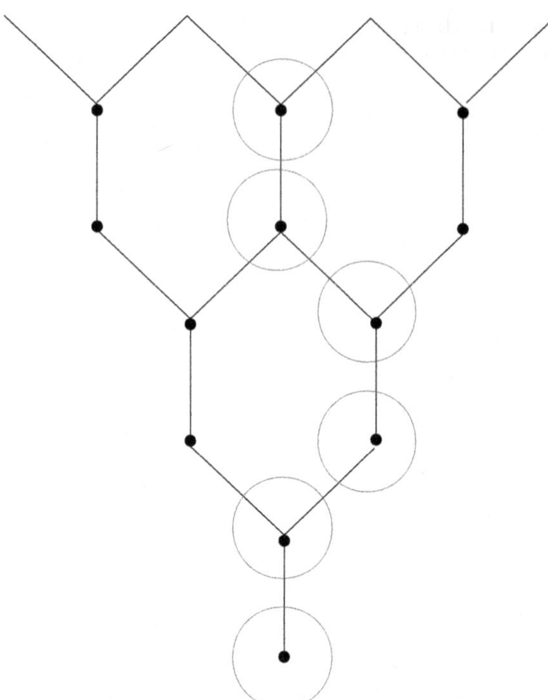

Fig. 7.14 Hexagonal configuration-c

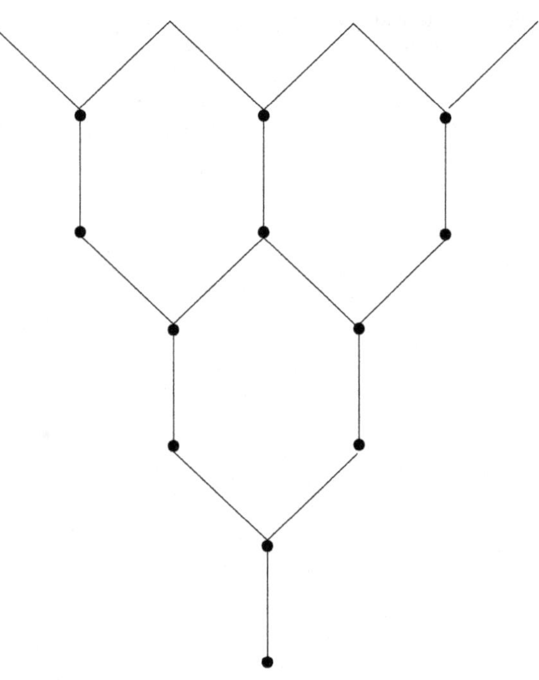

Fig. 7.15 Complete hexagonal configuration

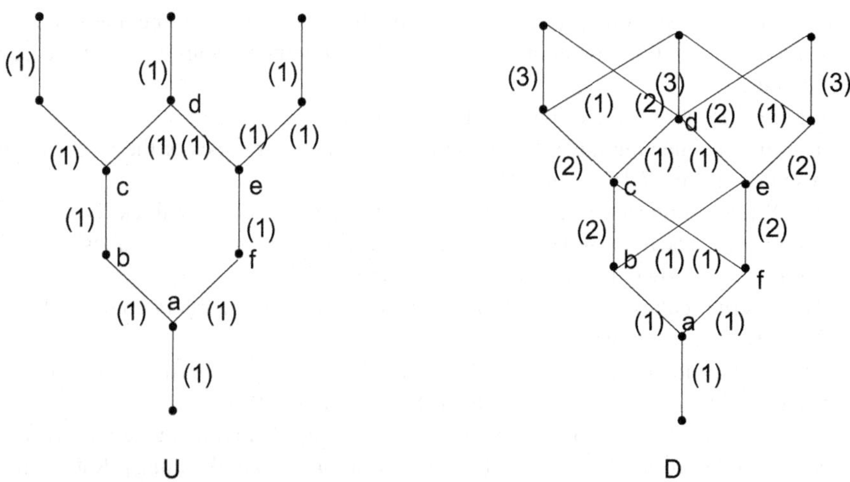

Fig. 7.16 U and D graphs for hexagonal configurations

Fig. 7.17 Commutation condition

U (b) = c

D (b) = a

D U (b) = 2 x b +f

U D (b) = b + f

(D U — U D) (b) = b

A lattice of Modular Self-Reconfigurable MSR, project consists of active links and passive cubes. The links are moveable and the cubes can be rotated, translated simultaneously in two directions, and act as a pivot joint for a moving link. The structures assembled from I-cubes are potentially able to move over obstacles, climb stairs, traverse through tunnels and pipes, manipulate objects, form bridges and towers, and be utilized for space applications (Unsal and Khosla 2001). Experiments using a physical prototype demonstrated basic link function and cube movement. A sequence of actions for climbing a step and building a tower were created with the aid of a graphical interface.

The connections are established by a cross or cone-shaped piece that locks into place. The modules are able to sense joint positions. The structures move by means of joint rotations of the cubes. The links are controlled externally by buttons or by a graphical user interface. Future work involves constructing smaller and lighter cubes, constructing more links and cubes, enabling the modules to be autonomous, and devising motion planning schemes that combine learning and search techniques.

Figure 7.18 shows the polytope 5D for cubes.

The lines connecting the 3D cubes, that is, the modular cubes or the I-cubes, are represented in Fig. 7.18 by thick hyper-lines.

The 8 lines connecting the two 3D cubes are represented by a thick hyper-line.

Using the thick hyper-lines, the 5D cube is shown as "square of cubes".

Figure 7.19 shows the polytope 6D for cubes.

The internal cubes again should have the corresponding corners connected.

The 6D cube is shown as "cube of cubes".

High-dimensional polytopes as shown in Fig. 7.19 have been investigated as parallel computers architectures (Seitz 1985; Ammon 1998).

According to Weller et al. (2011) we are on the verge of realizing new classes of materials that need not be machined or molded in order to make things. Rather, the materials form and re-form according to software programmed into component elements. These self-reconfiguring materials are composed of robotic modules that coordinate with each other locally to produce global behaviors. These robotic materials can be used to realize a new class of artifact, a shape that can change over time that is a 4D-shape.

The nD-forms, with n = 4, 5 or 6 present several opportunities. Objects such as furniture could exhibit dynamic behaviors, could respond to tangible and gestural input, and end-users could customize their form and behavior. The behaviors that nD-forms can perform will be constrained by the capabilities of the self-reconfiguring materials they are made of. By considering how we will interact with nD-forms, it is possible to inform the design of such systems (Weller et al. 2011).

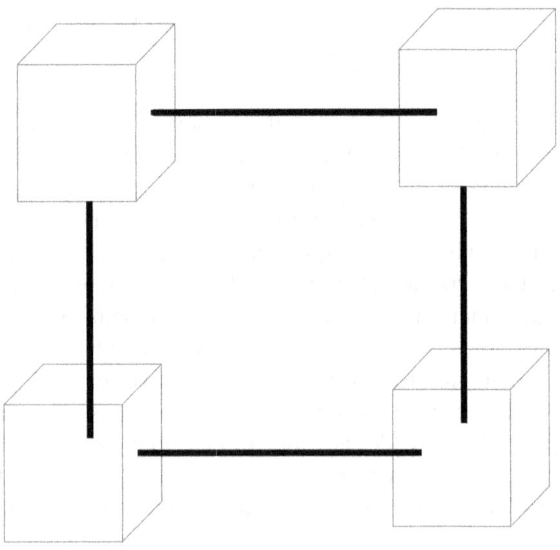

Fig. 7.18 Polytope 5D for cubes

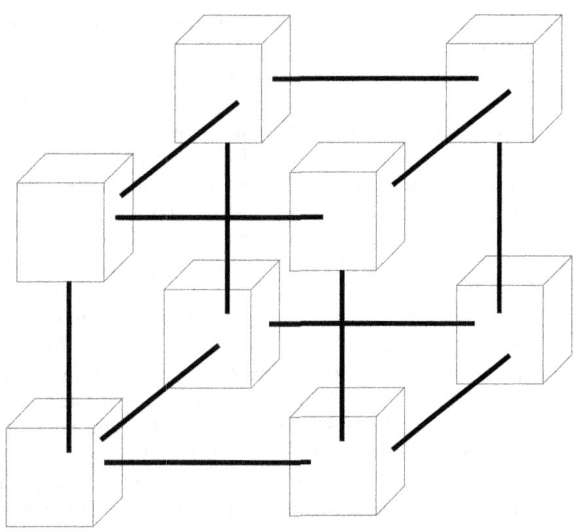

Fig. 7.19 Polytope 6D for cubes

References

Aloupis, G., Collette, S., Damian, M., Demaine, E.D., Flatland, R., Langerman, S., O'Rourke, J., Ramaswami, S., Sacristan, V., Wuhrer, S.: Linear reconfiguration of cube-style modular robots. Comput. Geom. Theory Appl. **42**, 652–663 (2009)

Ammon, J.: Hypercube Connectivity within ccNUMA Architectures. SGI Origin Team, Mountain View (1998)

Campbell, T.A., Tibbits, S., Garrett, B.: The Next Wave: 4D Printing. Atlantic (2014)

Christensen, D.J., Stoy, K.: Selecting a meta-module to shape-change the ATRON self-reconfigurable robot. In: Robotics and Automation, ICRA 2006. Proceedings 2006 IEEE International Conference on 2006 May 15, pp. 2532–2538. IEEE (2006)

Dorst, K.: Describing Design -a Comparison of Paradigms, Ph.D. thesis TUDelft (1997)

Ferreiro, V., Douglas, J.F., Warren, J., Karim, A.: Growth pulsations in symmetric dendritic crystallization in thin polymer blend films. Phys Rev E **65**(051606), 1–16 (2002)

Fomin, S.: Duality of graded graphs. J. Algebraic Combin. **3**, 357–404 (1994)

Forcey, S., Siehler, J., Seth Sowers, E.: Operads in iterated monoidal categories. J. Homotopy Relat. Struct. **2**, 1–43 (2007)

Gilpin, K., Rus, D.: Modular robot systems: from self-assembly to self-disassembly. IEEE Robot. Autom. Mag. **17**(3), 38–53 (2010)

Gouyou-Beauchamps, D., Nadeau, P.: Signed enumeration of ribbon tableaux with local rules and generalization of the Schensted correspondence. In: Proceedings of 19th International Conference on Formal Power Series & Algebraic Combinatorics, 2–6 July 2007, Nankai University, Tianjin, China (2007)

Iordache, O.: Self-evolvable Systems. Machine Learning in Social Media. Springer, Berlin (2012)

Jorgensen, M.W., Ostergaard, E.H., Lund, H.H.: Modular ATRON: modules for a self-reconfigurable robot. In: Intelligent Robots and Systems (IROS 2004). Proceedings. IEEE/RSJ International Conference on 2004 Sept 30, vol. 2, pp. 2068–2073. IEEE (2004)

LaCombe, J.C., Koss, M.B., Frei, J.E., Giummarra, C., Lupulescu, A.O., Glicksman, M.E.: Evidence for tip velocity oscillations in dendritic solidification. Phys. Rev. E **65**(031604), 1–6 (2002)

Lam, T.: Signed differential posets and sign-imbalance. J. Combin. Theory Ser. A **115**, 466–484 (2008)
Lam, T.: Quantized dual graded graphs. Electron. J. Combin. **17**(1), Research Paper 88, (2010)
Mahalik, N.P.: Micromanufacturing and Nanotechnology. Springer, New York (2005)
Propp, J.: Some variants of Ferrers diagrams. J. Comb. Theory Ser. A. **52**(1), 98–128 (1989)
Rus, D., Vona, M.: A basis for self-reconfiguring robots using crystal modules. In: Intelligent Robots and Systems, (IROS 2000). Proceedings. 2000 IEEE/RSJ International Conference on 2000, vol. 3, pp. 2194–2202. IEEE (2000)
Seitz, C.L.: The cosmic cube. Commun. ACM **28**(1), 22–33 (1985)
Stanley, R.P.: Differential posets. J. Amer. Math. Soc. **1**, 919–961 (1988)
Stucker, B.: Additive manufacturing technologies: technology introduction and business implications. In: Frontiers of Engineering: Reports on Leading-Edge Engineering From the 2011 Symposium, pp. 19–21. National Academies Press, Washington, D.C. (2012)
Tibbits, S.: 4D printing: multi-material shape change. Architectural Des. **84**(1), 116–121 (2014)
Unsal, C., Khosla, P.K.: A multi-layered planner for self-reconfiguration of a uniform group of i-cube modules. In: Intelligent Robots and Systems. Proceedings. 2001 IEEE/RSJ International Conference on 2001, vol. 1, pp. 598–605. IEEE (2001)
Weller, M.P., Gross, M.D., Goldstein, S.C.: Hyperform specification: designing and interacting with self-reconfiguring materials. Pers. Ubiquit. Comput. **15**(2), 133–149 (2011)

Chapter 8
Prospection and Retrospection

8.1 Smart Enterprises

8.1.1 Smart Factory

More and more things become smart and people strive for developing not only new and innovative devices, but also homes, factories, cities or even societies. Despite of continuous development, many of those concepts are still being just a vision of the future, which still needs a lot of effort to define and to become true.

Features that are desirable for the smart factory would relate to being flexible and reconfigurable, adaptive or transformable, agile and lean. One of the ways to achieve some of those functionalities would be to apply modular structure with respect to product, process technology and organization.

A smart factory is a manufacturing solution that provides such flexible and adaptive production processes that will solve problems arising on a production facility with dynamic and rapidly changing boundary conditions in an environment of increasing complexity. This special solution could on the one hand be related to automation, understood as a combination of software, hardware and mechanics, which should lead to optimization of manufacturing resulting in reduction of unnecessary labor and waste of resource.

On the other hand, it could be seen in a perspective of collaboration between different industrial and nonindustrial partners, where the smartness comes from forming a dynamic organization.

The smart factory is a factory that context-aware assists people and machines in execution of their tasks (Zuehlke 2008; Kagermann et al. 2013).

Based on the definitions given for CPS (Cyber Physical Systems) and the IoT (Internet of Things), the smart factory can be considered as a factory where CPS communicate over the IoT and assist people and machines in the execution of their tasks (Baheti and Gill 2011).

O. Iordache, *Advanced Polytopic Projects*, Lecture Notes in Intelligent Transportation and Infrastructure,
https://doi.org/10.1007/978-3-030-01243-4_8

One of the main properties of this CPS is the advanced networking of the production systems using Internet standards. It is not only the machines that are communicating with each other-the workpieces are communicating more and more with the production technology. To do this, resources and workpieces have an identity in the IoT. The de-centrally organized production units have a distinct level of agility.

The interaction between workpieces and production technology in smart factories allows flexible and application-based reconfiguration of production systems.

A key factor for the success of the smart factory concept is the integrative development of products and production systems. This means that the interdisciplinary collaboration, from the product development process to the development of the corresponding production technology, must be raised to a new level within the company.

Agile production systems require equally agile software systems for the planning, simulation and control of manufacturing processes (Internet of Services).

Centralist concepts will be complemented with smart, high-dimensional de-centralized systems as a result of the gradual development of CPS.

The manufacturing industry has been facing several challenges, including sustainability and performance of production. These challenges are sourced from numerous factors such as an aging workforce, changes in the landscape of global manufacturing and slow adaption of smart manufacturing by implementing IT in manufacturing process.

In recent years, different governments have established initiatives to accelerate the use of the Internet of Things (IoT) and smart analytics technologies in the manufacturing industries and, consequently, to improve the overall performance, quality, and controllability of manufacturing process.

From 2011, the manufacturing sector showed an amount of interest in the new conception of smart factory, mainly in Germany (Kagermann et al. 2011, 2013). A plan developed under the auspices of the German Federal Government's High-Tech Strategy is outlined to be the framework of the fourth industrial revolution.

The first industrial revolution was the mechanization of production using water and steam power. Innovations related to coal powered engines, steam, cotton, steel, and railways helped to give us the first industrial revolution of mass production and mechanization.

It was followed by the 2nd industrial revolution which introduced mass production mainly with the help of electric power. This revolutionary step was triggered by the introduction of electricity, heavy and mechanical engineering and synthetic chemistry.

This was followed by the 3rd revolution, the use of electronics and information technology to further automate production. The 3rd industrial revolution was triggered by innovations in electronics and computers, materials for electronics and aerospace, petrochemicals.

The 4th industrial revolution entails the fusion of digital with the physical industrial production in cyber-physical systems. It is characterized by a synthesis of technologies that is suppressing the lines between the physical, biological and digital levels. A host of new technologies are driving a wave of innovation that takes us into the new stage of industrial revolution. This concerns the internet, nanotechnology, bioscience, electronics, photonics, advanced and smart materials and renewable energies. Changes to the techno-economic system started in the mid-1980s, but we had to wait the turn of the century to witness their impact on the production methods (Schwab 2016; Hwang 2016).

Industry 4.0 initiative is influencing thinking throughout the world, which in turn influences other initiatives and cooperative efforts.

The term Industrie 4.0 originated in Germany, but the concepts are in harmony with worldwide initiatives. Similar digital manufacturing initiative appeared in different countries.

The Made in China 2025 strategy is part of a larger modernization campaign of China. The core elements of the Chinese strategy are enhanced creative ability, improved quality and efficiency, as well as green development. Until 2025 industry and information technology are supposed to be integrated and China's capacity to innovate and manufacturing productivity are planned to have been improved.

The Japanese Industrial Value Chain Initiative (IVI) is a platform to combine manufacturing and information technologies and facilitate collaboration between companies. It was founded in 2015 and aims to discuss potentials of human-centric manufacturing processes and to build a mutually connected system architecture (Wang et al. 2017).

On the other hand, the USA as a worldwide pioneer in manufacturing and smart systems developed and implemented concepts as smart manufacturing and cyber-physical systems (CPS) (Baheti and Gill 2011).

The Smart Manufacturing Leadership Coalition (SMLC) was founded in the USA to overcome the costs and risks associated with the commercialization of smart manufacturing systems. The SMLC has not explicitly embraced Industry 4.0, but its vision and mission embrace many of similar concepts (Bryner 2012; Davis et al. 2012).

Germany has enacted its Industrie 4.0 program, which is increasingly affecting European policy while USA focuses on smart manufacturing as global program (Thoben et al. 2017).

Smart manufacturing was defined as a data intensive application of information technology at the shop floor level and above to enable intelligent, efficient, and responsive operations.

Smart manufacturing incorporate many technologies including but not limited to CPS.

CPS is a complex engineering system that integrates physical, computation and networking, and communication processes. CPS can be illustrated as a physical device, object, equipment that is translated into cyberspace as a virtual model. With networking capabilities, the virtual model can monitor and control its physical aspect, while the physical aspect sends data to update its virtual model. Considering the importance of this topic, cyber-physical, Lee et al. (2015) presented a cyber-physical systems architecture for industry 4.0-based manufacturing systems. The main levels and sub-levels are as follows:

- Connection

  Plug & Play
  Free communication
  Sensor network.

- Conversion

  Component machine health
  Data correlation
  Performance prediction.

- Cyber-Physical

  Model for components and machines
  Clustering for similarity in data mining.

- Cognition

  Simulation and synthesis
  Diagnostics and decision making.

- Configuration

  Self-configure for resilience
  Self-adjust for variation
  Self-optimize for disturbance.

Figure 8.1 shows the polytope 4D associated to a cyber physical system The notations are:

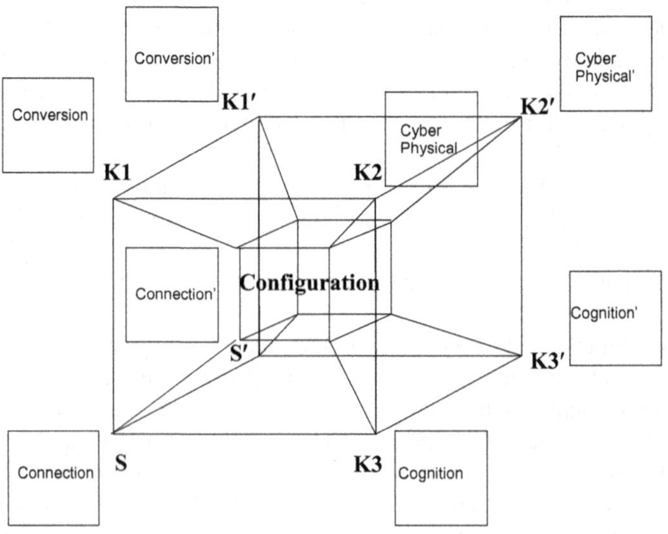

Fig. 8.1 Polytope 4D for cyber physical system

- S-Connection
- K1-Conversion
- K2-Cyber Physical
- K3-Cognition.

The "Self "was identified as Configuration, for this polytope.

Figure 8.2 shows polytope 4D for smart factory.

The notations are:

- S-Field
- K1-Control
- K2-Process Control
- K3-Plant Manage.

The "Self" is identified as ERP.(enterprise resource planning).

In Fig. 8.2 the traditional automation pyramid for manufacturing is presented differently. This automation pyramid handles both the product and the supply chain of the manufacturing at various stages. There are mainly five levels in the automation pyramid: company level at the top layer, ERP, plant level with the manufacturing execution systems K3, process level with all the material flow management computers K2, control level with the programmable logic controllers K1, and the field level with the physical devices S.

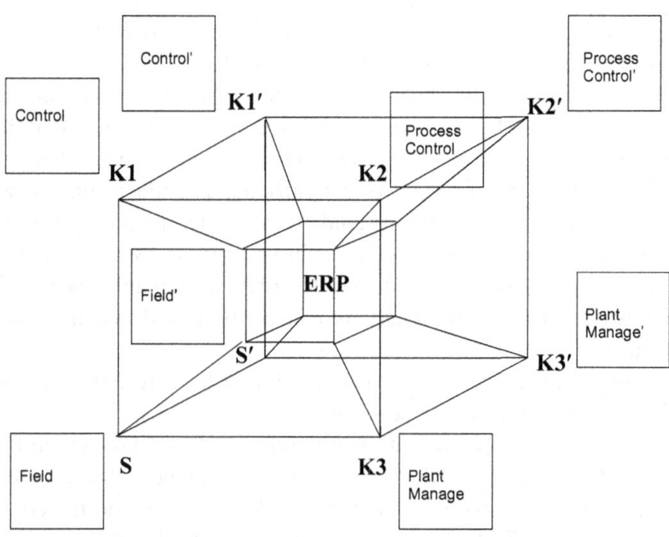

Fig. 8.2 Polytope 4D for smart factory

8.1.2 Reference Architectures

The Industrial Internet is an internet of things, machines, computers and people enabling intelligent industrial operations using advanced data analytics for transformational business outcomes, and the Industrial Internet Consortium (IIC) is devoted making the Industrial Internet a reality.

The reference architecture addresses the Industrial Internet problems by providing common and consistent definitions in the system of interest, decompositions, and design patterns, and a common terminology to discuss the specification of implementations so that options may be compared.

Industry 4.0 and related initiatives recognize that efficiently building self-managing production processes requires open software and communications standards that allow sensors, controllers, people, machines, equipment, logistics systems, and products to communicate and cooperate with each other directly. Future automation systems must adopt open source interoperability software application and communication standards similar to those that exist for computers, the internet, and smart phones.

Industry 4.0 demonstrations acknowledge this by leveraging existing standards, including the ISA-88 batch standards, ISA-95 enterprise-control systems integration standards, IEC 6-1131-3, and others standards.

The harmonization of standards worldwide recently took another step forward when providers of the Platform Industry 4.0 and the Industrial Internet Consortium (IIC) met to explore the potential alignment of their two architecture efforts-respectively, the Reference Architecture Model for Industry 4.0 (RAMI4.0) and the Industrial Internet Reference Architecture (IIRA).

The OPC Foundation and Object Management Group (OMG) initiated a collaborative strategy for technical interoperability that encompasses the OPC Unified Architecture (OPC UA) and the OMG Data Distribution Service (DDS) standards.

These significant cooperative efforts recognize that manufacturing has worldwide interdependencies requiring common standards and interoperability (OMG 2008).

The Reference Architectural Model Industry 4.0, RAMI 4.0, consists of a three-dimensional coordinate system that describes all crucial aspects of industry 4.0. In this way, complex interrelations can be broken down into smaller and simpler clusters.

The Hierarchy Levels axis is based on the levels from IEC 62264, the international standards series for enterprise IT and control systems.

These Hierarchy Levels represent the different functionalities within factories or facilities. In order to represent the industry 4.0 environment, these functionalities have been expanded to include workpieces, labeled Product, and the connection to the Internet of Things and Services, labeled Connected World.

The Life Cycle and Value Stream axis represents the life cycle of facilities and products, based on IEC 62890 for life-cycle management. Furthermore, a distinction is made between types and instances. A type becomes an instance when design and prototyping have been completed and the actual product is being manufactured.

RAMI 4.0 combine the basic elements of industry 4.0 in a three-dimensional layer model. Based on this framework, different industry 4.0 technologies can be classified and further developed.

The Layers axis show six layers on the vertical axis and serve to describe the decomposition of a machine into its properties structured layer by layer, that is, the virtual mapping of a machine. Such representations originate from information and communication technology, where properties of complex systems are commonly broken down into layers.

These six layers of the Layers axis are:

- Asset: representation of reality, such as technical subjects and resources
- Integration: providing computer processing information of assets
- Communication: standardization of communication, using a unified data format
- Information: software environment for event pre-processing
- Functional: modeling environment for services that support business processes
- Business: business models and the resulting business process.

Within these three axes, all crucial aspects of industry 4.0 can be mapped, allowing objects such as machines, technologies, test-beds to be classified according to the model.

Highly flexible industry 4.0 concepts can thus be described and implemented using RAMI 4.0. The reference architectural model allows for step-by step migration from the present industrial stage into the world of industry 4.0.

RAMI 4.0 integrate different user perspectives and provide a common understanding of industry 4.0 technologies. With RAMI 4.0, requirements of sectors from manufacturing automation and mechanical engineering or chemical process engineering can be addressed in standardization committees. Thus, RAMI 4.0 provide a common understanding for standards and use cases.

RAMI 4.0 can be regarded as a 3D map of industry 4.0 solutions. It provides an orientation for plotting the requirements of sectors together with national and international standards in order to define and further develop industry 4.0.

8.1.3 Generic Smart Grid Architecture Model

The generic Smart Grid Architecture Model, SGAM, can act as a reference designation system in order to describe smart grid technical use cases as well as business cases.

The approach used in SGAM for reference designation proved its value, and it is necessary to follow this basic guideline for successful adoption of derived models for other domains (Fang et al. 2012; Uslar and Engel 2015).

One of the key challenges resulting from the Smart Grid vision is to handle complexity in the new distributed systems landscape. The Smart Grid, being a real system-of-systems is a prime example for the increasing complexity that emerges in any distributed system.

SGAM provides the means to express various domain-specific viewpoints on architecture models by the concepts of so called Zones, Domains and Layers.

Figure 8.3 shows the polytope 5D for SGAM.

Using the thick hyper-lines, the 5D cube is shown as "square of cubes".

The selected basic levels are Assets, Zones, Domains, and Layers.

Asset refers to representation of reality and shows resources available.

Details for levels will be presented in the following.

Figure 8.4 shows the polytope 4D for SGAM Layers.

Figure 8.4 shows the Layers for the original SGAM model for reference designation of standards.

The elements of the polytopic architecture from Fig. 8.4 have been identified as follows:

- S-Component
- K1-Communication
- K2-Information
- K3-Function
- Self-Business.

Business is identified as the "Self" of this polytope.

The Domains of SGAM regard the energy conversion chain and include: Generation (conventional and renewable bulk generation capacities), Transmission (infrastructure and organization for the transport of electricity), Distribution (infrastructure and organization for the distribution of electricity), DER (distributed energy resources connected to the distribution grid) and Customer Premises (both end users and producers of electricity, including industrial, commercial, and home facilities as well as generation).

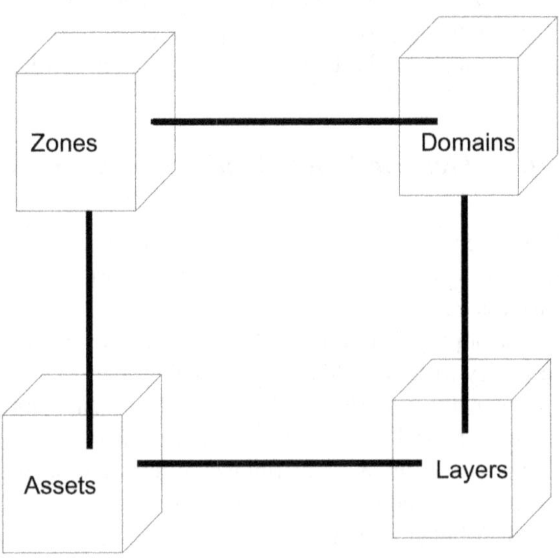

Fig. 8.3 Polytope 5D for SGAM

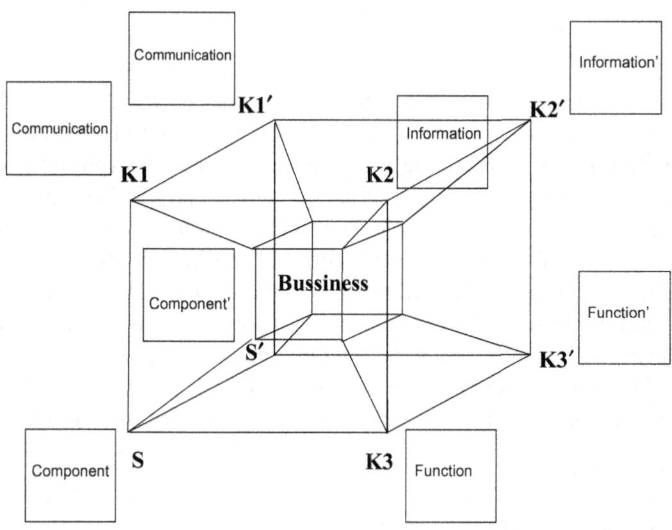

Fig. 8.4 Polytope 4D for SGAM Layers

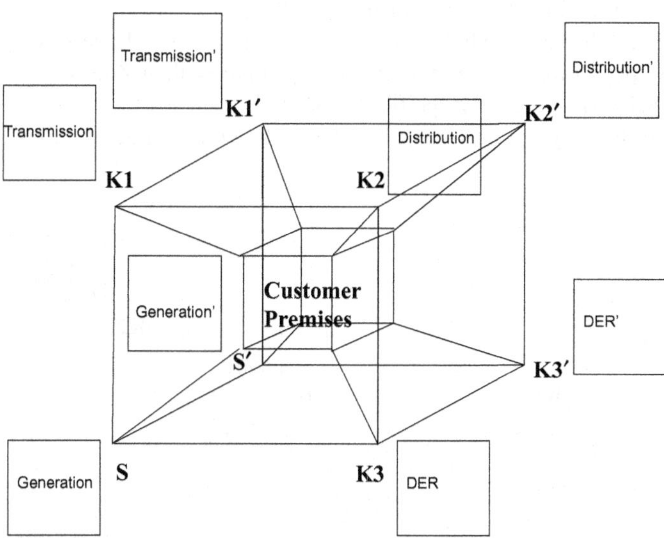

Fig. 8.5 Polytope 4D for SGAM Domains

Figure 8.5 shows the Domains for the original SGAM.

The elements of the polytopic architecture are identified as follows:

- S-Generation
- K1-Transmission
- K2-Distribution

- K3-DER
- Self-Customer Premises.

Customer Premises are identified as the "Self" of this polytope.

The hierarchy of power system management from the automation perspective is reflected within the SGAM by the following Zones: Process (physical, chemical, biological or spatial transformations of energy and the physical equipment directly involved), Field (equipment to protect, control and monitor the process of the power system), Station (areal aggregation level for field level), Operation Control (power system control operation in the respective domain), Enterprise (commercial and organizational processes, services and infrastructures for enterprises), and Market (market operations possible along the energy conversion chain).

Figure 8.6 shows the Zones for the original SGAM.

The elements of the polytopic architecture are identified as follows:

- S-Process Field
- K1-Station
- K2-Operation Control
- K3-Enterprise
- Self-Market.

Market is identified as the "Self" of this polytope.

SGAM may be represented using multi-level polytopes. Layers covers coarser granularity that Domains and these coarser granularity than Zones.

Figure 8.7 shows the polytope 6D for SGAM.

With the thick hyper-lines, the 6D cube is shown as "cube of cubes".

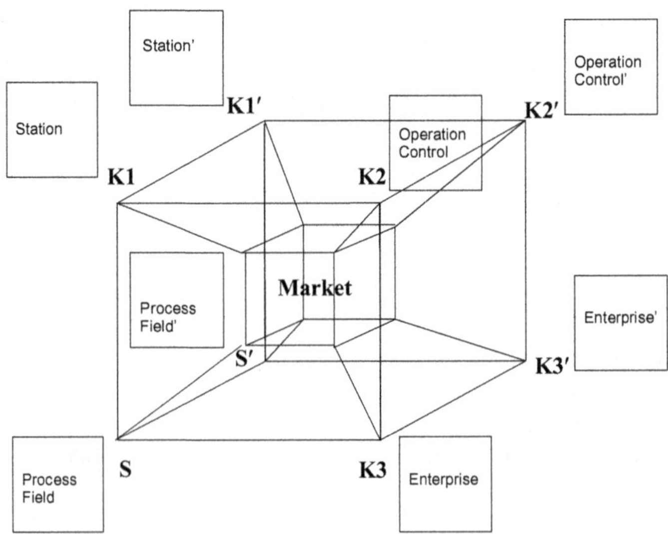

Fig. 8.6 Polytope 4D for SGAM Zones

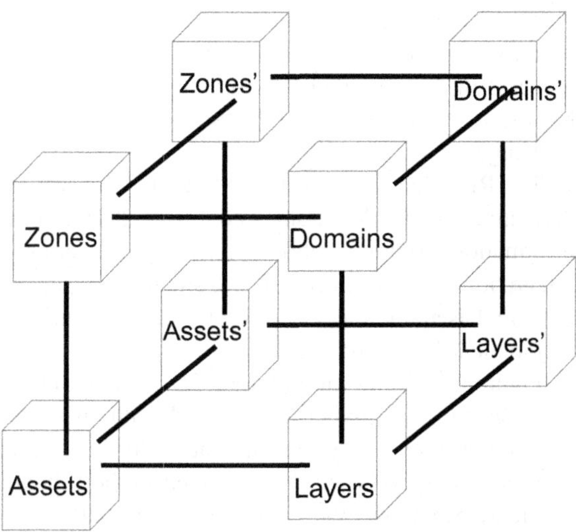

Fig. 8.7 Polytope 6D for SGAM

The basic levels: Assets, Zones, Domains, and Layers are modified to the basic levels Asset', Zones', Domains', and Layers'.

Reference architectures based on SGAM as generic reference architecture have been presented by Uslar and coworkers (Uslar and Trefke 2014; Uslar and Engel 2015; Uslar and Gottschalk 2015).

The Smart City Infrastructure Architecture Model (SCIAM) is one particular new derivative from the original SGAM model. The roadmap for Smart Cities is based on the original model of the SGAM. Instead of the business layer, an action layer was proposed.

As for Domains and Zones, new axes have been developed.

The Zones cover a mostly hierarchical way of structuring for physical locations. Market, Enterprise, Operation, Station and Field as well as Process, forms the Zones axis. This list can be considered a natural ordered list. In addition to this, the Domains consist of Supply/Waste Management, Water/Waste Water, Mobility and Transport, Healthcare and Civil Security, Energy, Buildings as well as Industry.

The Electric Mobility Architecture Model (EMAM) is a particular architecture which is currently being developed in the context of the IT for electric vehicles programs (Uslar and Gottschalk 2015).

It is a need for a consolidated use case collection and then deriving actors and technical requirements from them which will provide the basis of changing the granularity of the individual axis aspects. Re-using the SGAM in terms of modeling electric mobility is required.

The concept of the Home and Building Architecture Model (HBAM) has been developed to come up with a Standardization Roadmap on Smart Home and Building.

The Layers have been renamed to application, function, data model, interface and protocol and finally component.

The Zones axis contains the electronic health, building automation, physical security, consumer electronics and energy domain. Just like with the SCIAM more domains than one are addressed, but this time in the Zones area. The Domain axis has been structured with the lanes of devices, interfaces, control, accesses and data exchange.

The Reference Architecture Model for Industry 4.0 (RAMI 4.0) is an advanced derivative of SGAM (Uslar and Engel 2015). In addition to business, function, information, communication and asset representing component, a new layer called integration was introduced. The Domain and Zone axis are not custom taxonomies but are based on the IEC 62890 value stream chain or the IEC 62264/61512 hierarchical levels, respectively.

Figure 8.8 shows the polytope 5D for RAMI4.0.

Using the thick hyper-lines, the 5D cube is shown as "square of cubes".

The basic levels are: Assets, Hierarchy Levels, Value Streams and Layers.

Details for levels will be presented in the following.

Figure 8.9 shows the Layers for RAMI 4.0.

The elements of the polytopic architecture for Layers are identified as follows:

- S-Asset Integration
- K1-Communication
- K2-Information
- K3-Function
- Self-Business.

Business is identified as the "Self" of this polytope.

For this polytope presentation the asset and integration are considered together as a single layer.

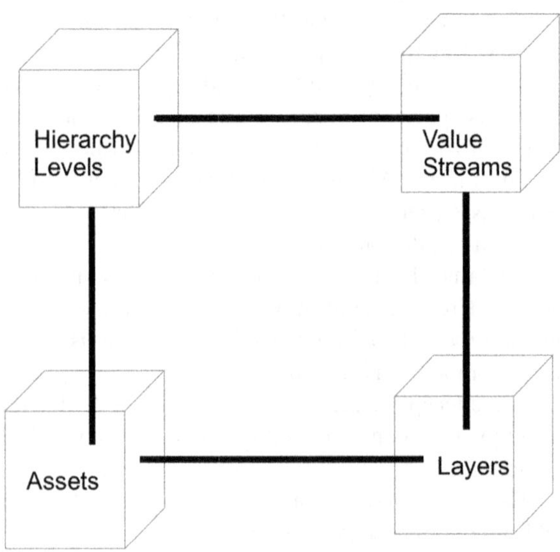

Fig. 8.8 Polytope 5D for RAMI4.0

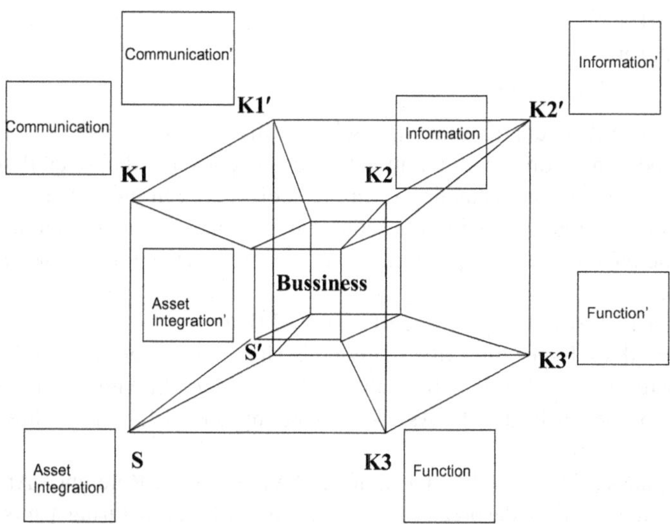

Fig. 8.9 Polytope 4D for RAMI 4.0 Layers

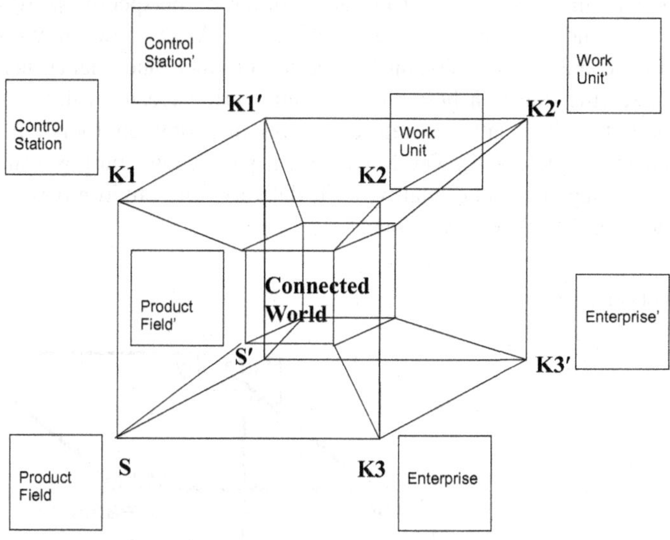

Fig. 8.10 Polytope 4D for RAMI 4.0 Levels

Figure 8.10 shows the Levels for RAMI 4.0.

The elements of the polytopic architecture from Fig. 8.10 are identified as follows:

- S-Product Field
- K1-Control Station

- K2-Work Unit
- K3-Enterpise
- Self-Connected World.

Connected world is identified as the "Self" of this polytope.

The model harmonizes different user perspectives on the overall topic and provides a common understanding of the relations between individual components for industry 4.0 solutions. Different industrial branches like automation, engineering and chemical process engineering have a common view on the overall systems landscape.

Figure 8.11 shows the polytope 6D for RAMI4.0.

With the thick hyper-lines, the 6D cube is shown as "cube of cubes".

The main levels Assets, Hierarchy Levels, Value Streams and Layers are modified to the main levels Assets', Hierarchy Levels', Value Streams' and Layers'.

The Japanese IVI published the Industrial Value Chain Reference Architecture (IVRA) to foster the widespread of so called Smart Manufacturing Units (SMUs), which are defined as individual units within industrial systems that interact with each other autonomously through mutual communication and thereby improve productivity and efficiency (Wang et al. 2017).

The model analyzes SMUs from three different perspectives: Asset view, Activity view, and Management view. While the Asset view shows resources available to the enterprise (Personnel, Process, Product, and Plant), the Activity view addresses the question how smart manufacturing creates values as the outcome of activities. The latter therefore deploys the common four-steps Deming cycle (Plan, Do, Check, Act). The Management view shows purposes and indices relevant for management like quality, cost, delivery, and environment, which are used for steering Assets and Activities.

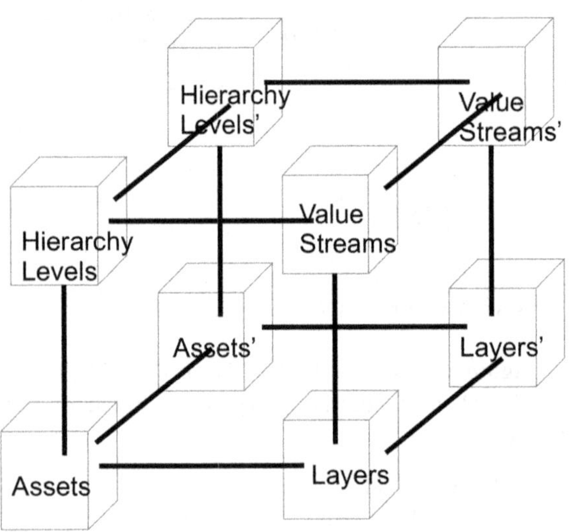

Fig. 8.11 Polytope 6D for RAMI4.0

Figure 8.12 shows the polytope 5D for IVRA.
Using the thick hyper-lines, the 5D cube is shown as "square of cubes".
The main levels are: Assets, Hierarchy Levels, Demand Supply and Layers.
Details for Levels will be presented in the following.
Figure 8.13 shows the polytope 4D for IVRA Levels.

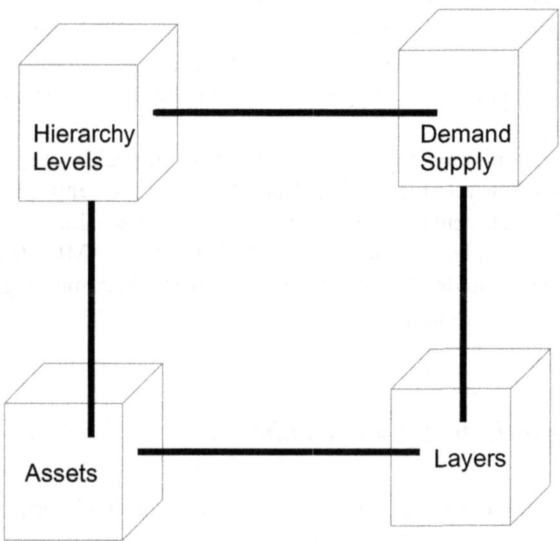

Fig. 8.12 Polytope 5D for IVRA

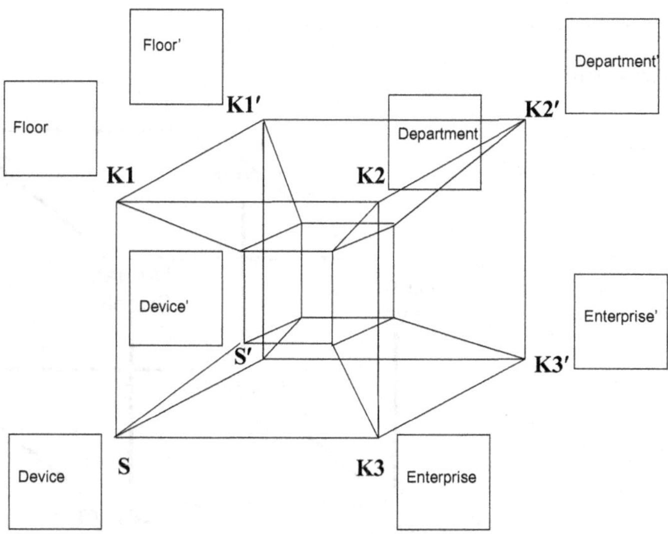

Fig. 8.13 Polytope 4D for IVRA Levels

In this case the notations are:

- S-Device
- K1-Floor
- K2-Department
- K3-Enterpise.

Figure 8.14 shows the polytope 6D for IVRA.

With the thick hyper-lines, the 6D cube is shown as "cube of cubes".

The main levels Assets, Hierarchy Levels, Demand Supply and Layers are modified to the main levels Assets', Hierarchy Levels', Demand Supply' and Layers'.

This modification may be due to a modified strategies or programs. These programs are not illustrated in Fig. 8.14 but they play the central role in transition from 5D to 6D. Such centers allow to increases the dimension.

Observe that the reference architecture for SGAM, RAMI 4.0 and IVRA are similar but do not coincide. This illustrates the underlying common ground between different digitalization initiatives.

8.1.4 Generic Industry 4.0 Toolbox

Industry 4.0 is increasingly entering the industrial landscape of Small and Medium-sized Enterprises (SME). However, the right introduction to the subject for the affected enterprises proves to be difficult. For this reason, a generic procedure model to introduce Industry 4.0 in SME was developed.

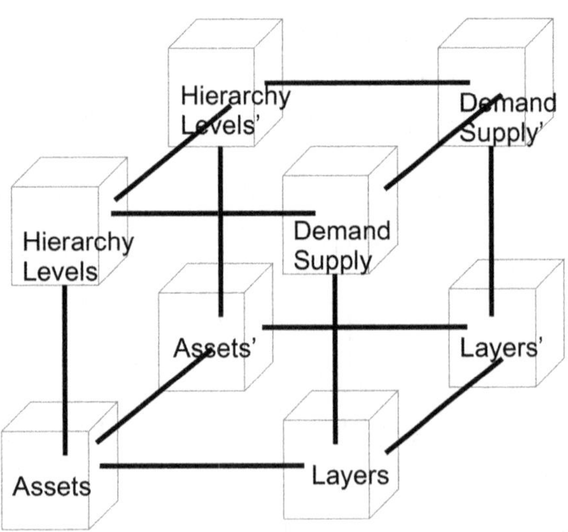

Fig. 8.14 Polytope 6D for IVRA

The purpose of this model is to give SMEs guidance on how to address Industry 4.0 in order to identify enterprise-specific technology solutions, to optimize existing processes, to advance existing business models and to exploit new business models (Anderl et al. 2015).

The development of toolboxes for engineering follows the Guideline Industry 4.0 set by the VDMA (*Verband Deutscher Maschinen- und Anlagenbau*).

To meet the guidelines, applications are firstly defined and divided in development stages. Relevant applications are identified based on the defined engineering tasks and potentials. Furthermore, the applications are classified respecting the categories of engineering potentials. The most important applications have to be selected and used in the toolbox.

Toolboxes are structured in applications vertically and development stages horizontally.

Application displays Industry 4.0 themes, where every application is broken down in five technological and sequential stages. The lowest level 0.0 correspond to the conventional usually non-digitalized stage. The highest level 4.0 represents the Industry 4.0 perspective. Toolboxes for Product, Production, Intra-logistics, Assembly and Security have been developed.

Application areas and performance levels are derived from the basic approach for Industry 4.0 as defined by the RAMI 4.0. The RAMI 4.0 terminology, however, has been mapped into a terminology which is used in industrial product development and industrial production. Application areas describe topics of interest in the context of Industry 4.0, for industrial product development as well as for industrial production. Performance levels describe technological approaches which are structured into a sequence reaching from none or low level industry performance labeled 0.0 to high industry 4.0 performance.

Table 8.1 shows examples of Industry 4.0 Toolboxes.

The stages may be numbered from 1 to 5 or 0.0 to 4.0.

Figure 8.15 shows these stages and proposes common illustrations for different applications.

The first column from Table 8.1 corresponds to the applications.

Consider that the application is the integration of sensors and actuators. The development stage starts with the product having no use of sensors and actuators (stage 0.0) and the Industrie 4.0 vision is that the product responds independently based on gained data (stage 4.0). The intermediary stages (1.0) (2.0) and (3.0) for any application are established based on the expertise in the domain. Expertise allows taking into account the previous qualitative progress steps for the specific domain of application.

Another application in Table 8.1 is the communication and connectivity. The development starts with the product having no interfaces (stage 0.0) and the Industrie 4.0 vision is that the product having access to the internet (stage 4.0).

Another example of application is the monitoring. The development starts with no monitoring by the product (stage 0.0) and the Industrie 4.0 vision is the independently adopted control measures of the product (stage 4.0).

Table 8.1 Industry 4.0 Toolbox

Application	0.0	1.0	2.0	3.0	4.0
Integration Sensors/ Actuators	No Sens/Act	Sens/Act Integrated	Sensors reading processed by system	Data evaluated analyzed by system	System independently respond based on data analysis
Communication Connectivity	No interface	I/O signals	Field bus interface	Ethernet interface	Product access to internet
Monitoring	No monitoring	Failure detection	Recording operating conditions	Prognosis of condition	Independent control measures
Product related IT service	No services	On line portals	Service direct via product	Independent performed service	Complete integration in infrastructure of IT services
Data processing	No processing	Storage data	Analyze data	Evaluation for planning control	Automatic process planning control
Company-wide networking	No networking	Info via mail phone	Rule for data exchange	Interdivision inked data servers	Interdivision fully networked IT solutions
ICT infrastructure	Info via mail phone	Central data servers	Internet portals	Automated Info exchange	Suppliers/ customers fully integrated
Efficiency with small batches	Rigid production systems	Flexible production Identical parts	Flexible production Modular design	Component driven Flexible production	Component driven Modular production Value-adding networks

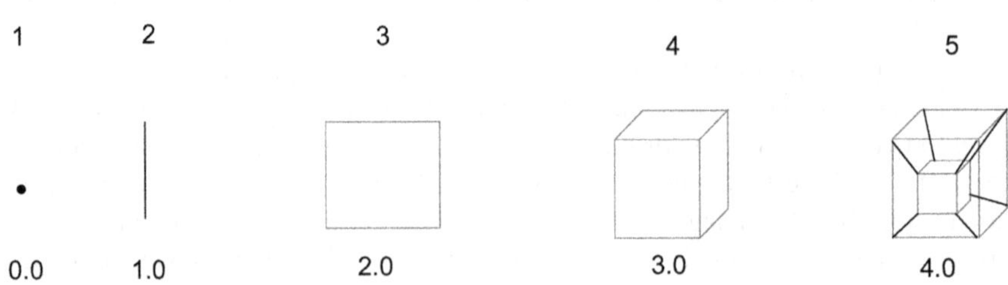

Fig. 8.15 Stages and illustrations

The next example of application is the product-related IT services. The development stage starts with no services and the Industrie 4.0 vision is that the complete integration into an infrastructure of IT services.

The stages 0.0, 1.0, 2.0, 3.0, 4.0 may be associated to S (K0), K1, K2, K3 and “Self” in polytopic architecture.

Figure 8.16 shows the polytope 4D for Industry 4.0 toolbox.

The stages 0.0’, 1.0’, 2.0’, 3.0’, 4.0’ may be associated to modified stages that is to S’ (K0’), K1’, K2’, K3’ and “Self’. The modifications means that once the application arrived at 4.0 all the preliminary stages should be modified (digitalized in many cases).

For instance, if we refer to monitoring, a stage like failure detection 1.0, should benefits from digitalizing and be modified at 1.0’.

A 5D polytope for Industry toolbox may be defined.

We can take into consideration the four applications referring to product:

- Integration
- Communication
- Monitoring
- IT Service.

They are the first four rows in the Table 8.1.

Figure 8.17 shows the polytope 5D for Industry 4.0 toolbox-product.

Using the thick hyper-lines, the 5D cube is shown as “square of cubes”.

The basic levels are Integration, Communication, Monitoring, and IT Service.

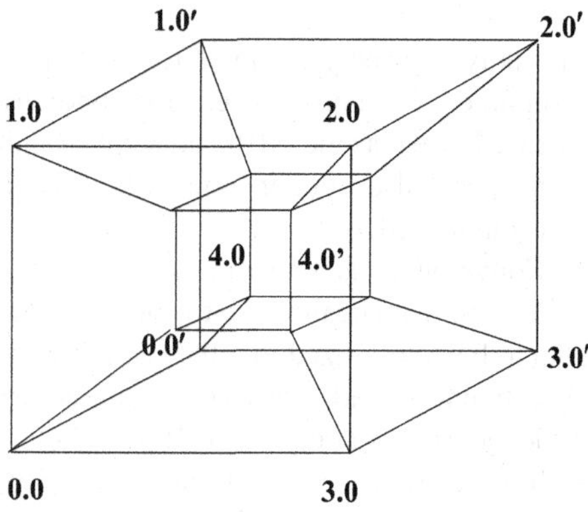

Fig. 8.16 Polytope 4D for Industry 4.0 toolbox

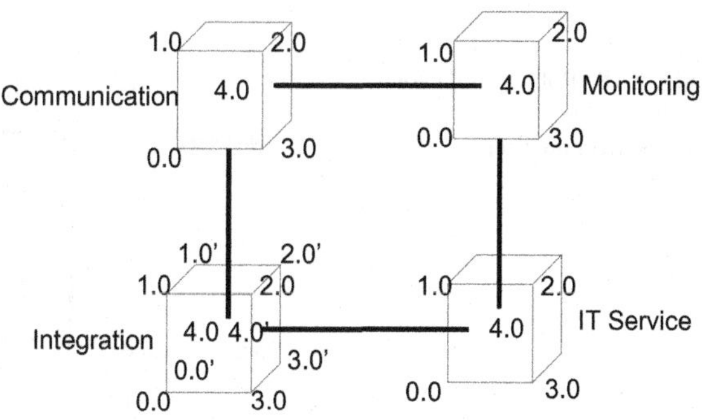

Fig. 8.17 Polytope 5D for Industry 4.0 toolbox-product

To any application domain we associate the stages 0.0 to 4.0 and the modified stages 0.0’ to 4.0’.

We can consider the four applications referring to production:

- Data Processing
- Networking
- ICT infrastructure
- Efficiency.

They are the next group of four rows in the Table 8.1.

Figure 8.18 shows the polytope 5D for Industry 4.0 toolbox-production.

Using the thick hyper-lines, the 5D cube is shown as “square of cubes”.

The developed 5D polytopes represents steps towards Industry 5.0.

The basic levels are Data Processing, Networking, ICT Infrastructure, and Efficiency.

8.2 Smart Technologies

8.2.1 NBIC

The polytopic projects joint recent trends advocating the convergence of several disciplines as, nanoscience, biotechnology, information technology and cognitive science known as the NBIC concept (Bainbridge and Roco 2006). This concept was developed also under the label BINC (bio-info-nano-cogno convergence) (Rasmussen 2016).

Convergence is a new paradigm that can yield critical advances in a broad array of sectors, from health care to energy, food, and climate (Sharp and Langer 2011). It should be emphasized that the convergence way does not quite grasp the essence of inventiveness and creativity required to confront high complexity problems. Tendencies to converge should be correlated with tendencies to diverge and it is the skillful relation of both trends in a polytopic architecture that matters. This is significant for technologies, data analysis, artifacts and for theories.

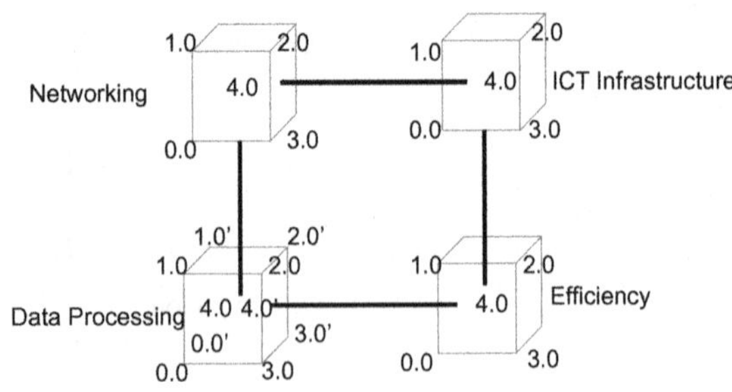

Fig. 8.18 Polytope 5D for Industry 4.0 toolbox-production

BINC approach may be seen also as an implementation of self-integrative closure concept were the convergence referring to matter in general is not restricted to nano domain (Iordache 2012, 2013, 2017).

The idea for integrative closure was to reassess and completes the interrelation between the four basic ontological levels in the study of nature: material, biological, cognitive or psychological and intelligent or logical (Hartmann 1952).

It was acknowledged that the hierarchical structures cannot serve as unique model for multi-level knowledge organization. Facing complexity the task of knowledge integration remains pertinent. Integrative closure is the strategy allowing confronting complexity.

The four basic levels of real systems have been denoted: S (or K0), K1, K2 and K3.

The integrative closure aims to make ends meets, for the four levels or realms, emphasizing the hypothetical interconnection between the material realm, S and intelligent realm, K3. The closed system is supposed to cross the gap between these farthest levels. This gap crossing corresponds to the cyber-physical systems. The cyber part is associated to K3 while the physical part is associated to S.

Integrative closure hypothesis is looking for structural analogies and a common methodology for different domains as bio-inspired systems, knowledge organization, problem solving or technological developments.

The growth in complexity is not only about differentiation but also about integration and coordination. The pursuit of evolvability towards further step of self-evolvability for systems imposed the study of polytopic architecture associated to the general framework as shown in Fig. 8.19. Figure 8.19 illustrates the concept of self-integrative closure with specific reference to the Hartmann's four levels of the real world.

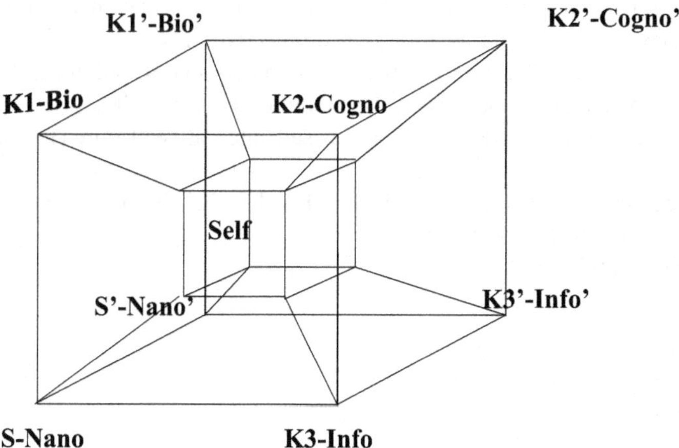

Fig. 8.19 Self-integrative closure for NBIC

The basic levels are S, K1, K2 and K3 represented on the front face of the outer cube and S′, K1′, K2′ and K3′ represented on the back face of the outer cube of the polytope.

Following Hartmann ontology, for the polytope shown in Fig. 8.19 we could identify:

- S-Nano Systems
- K1-Bio Systems
- K2-Cogno Systems
- K3-Info Systems.

Figure 8.19 shows the self-integrative closure for NBIC.

For technological and engineering systems we refer to a hierarchy of artificial systems inspired by the four level of Hartman's ontology as for instance:

- S-Materials or Smart Materials, Evolving or Living Technologies
- K1-Bio-Inspired or Bio-Mimetic Systems, Artificial Life
- K2-Cognitive-Inspired Systems, Neural Computer, Artificial Brain
- K3-Artifacts, Artificial Intelligent System, Cyber-Physical Systems.

The objects of engineering are natural systems and artifacts or artificial systems and that the reality levels as biological, cognitive and intelligent may serve for inspiration purposes.

For instance, evolutionary computation studies and evolvable devices may be inspired by biological principles but do not attempt to model or to mimic detailed data or processes from real genomes. Bio-inspired artificial design is not constrained by high fidelity to the original natural complex system. Examples include genetic algorithms calculus inspired by evolution and genetics, artificial neural networks and artificial neural codes inspired by cognition studies but not restricted to this, artificial intelligence and so on.

To re-apply the engineering understanding in developing new ways of study or explanations of biological, cognitive or intelligence of relevance for real world systems, may be considered as a long-term objective only.

The swinging between the two faces of the outer cube in polytope as shown in Fig. 8.19 is mediated by the inner cube identified as the "Self".

This reference framework was labeled as self-integrative closure (Iordache 2012, 2013).

Modifications of the direct sequence S, K1, K2, K3 during reverse processing K3′, K2′ K1′ and S′ may happen.

We may identify

- S′-Nano′
- K1'-Bio′
- K2′-Cogno′
- K3′-Info′.

Emergency from the direct sequence is completed by re-emergency in the reverse sequence. During these complementary processes of emergency, de-emergency, and re-emergency the differences allowing and accompanying self-evolvability and smartness may be generated.

The fourth industrial revolution may be considered as a more radical BINC (Rasmussen 2016). The first industrial revolution was the mechanization of production using water, wind and steam power (industry 1.0).

During the nineteenth century, the industrial revolution automated mass production in factories and a vast transportation infrastructure (industry 2.0). In the latter part of the twentieth century and the start of current century, the information technology revolution automated personal information processing in computers (industry 3.0). The next major technological revolution in part will be based on an integration of information processing and material production akin to living organisms and ecosystems seamlessly combining these processes (industry 4.0). The technological basis for the current transition has been named also the bio-info-nano-cogno convergence (BINC).

Major transitions in human societies are usually characterized by new tools and new infrastructures, new ways of making a living, changes in trade, new institutions and organizational forms, new local and global power structures as well as a new collective narrative (Rasmussen 2016). The "Self" of polytopic projects may be associated to narratives of BINC.

Globalization, mainly powered by information technology, is presenting us with a complex network of interconnected phenomena, including taxation and control issues for global companies, profit from digital products without production, which together with other issues cause the nation state to lose its influence over its own destiny. Transparency and privacy issues are increasingly taking the spotlight, because the new technologies generate new questions about who should have insight in what, and who own or has access to an individual's data.

The BINC or NBIC interface will soon enable individuals to design, transmit, produce and recycle most needed material goods as well as personalized medicine within the confines of their home through personal fabrication devices. This poses interesting opportunities for advances in regional sustainability as well as major challenges for current concepts in economy, future employment and ownership of the means of production.

The new opportunities and challenges we face through the BINC or NBIC technology convergence require that all components of society engage with these new technologies trends in a responsible manner and work together to develop a free and open society.

Figure 8.20 shows the polytope 5D for NBIC.

Using the thick hyper-lines, the 5D cube is shown as "square of cubes".

The basic realms or levels are: Nano, Bio, Cogno and Info.

Within each realm there is a recurring cycle of four sub-levels:

0-set
1-mapping (of sets)
2-system (composition of mappings)
3-system of systems.

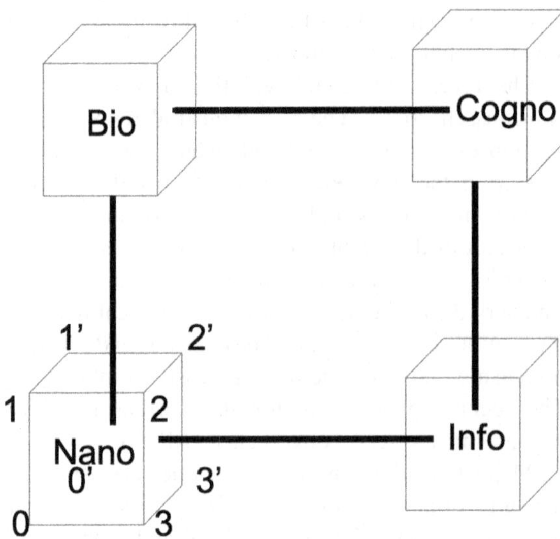

Fig. 8.20 Polytope 5D for NBIC

The decomposition in sub-levels as: set-0, mapping-1, system-2 and system of systems-3 follow suggestions from dynamic skill theory of development (Fischer 2008; Halford and Andrews 2010).

Figure 8.21 shows the polytope 6D for NBIC.

With the thick hyper-lines, the 6D cube is shown as "cube of cubes".

The basic realms are: Nano, Bio, Cogno and Info are modified to Nano', Bio', Cogno' and Info'.

This modification may be due to a modified strategy. This new strategy plays the central role in transition from 5D to 6D.

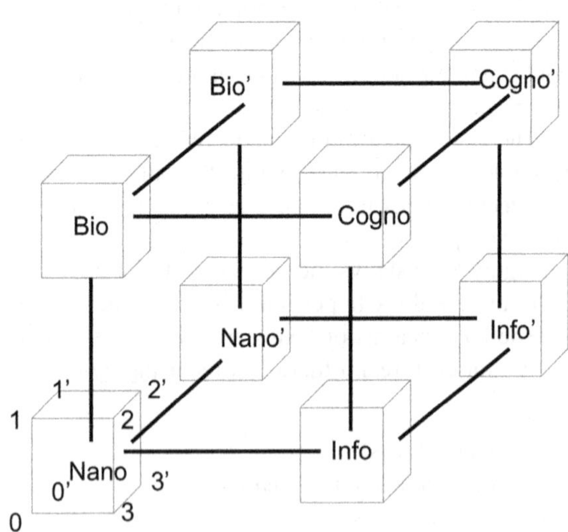

Fig. 8.21 Polytope 6D for NBIC

8.2.2 Leading Projects

The future leading technological and scientific projects require coding exploration and target artificial living, cognitive and intelligent systems.

Consistent with Hartmann ontology, we refer to the four basic levels of reality: material, biological, cognitive and intelligent. The associated abbreviations are: Mat, Bio, Cogno, and Intel. Due to their specific role, the mathematical objects pertaining to the intelligent level have been highlighted under the term Math. Math is not a separate level but is included in Intel.

Table 8.2 shows examples of leading projects for different levels of reality.

Coding research starts from the periodic table of elements and continues with Human Genome project and to BRAIN initiative (Jorgenson et al. 2015).

Living technologies, artificial life (von Kiedrowski et al. 2010), artificial brain, artificial intelligence and smart systems are leading scientific and technological projects.

Living technology is promising because it shares the fundamental properties of living systems (Bedau et al. 2010). These include self-assembly, self-organization, growth and division, purposeful action, adaptive complexity, evolution, and intelligence. Existing technologies are becoming increasingly like-alive, and powerful. Examples of living technology projects are synthetic biology attempts to make living systems from scratch in the laboratory, systems exhibiting collective and swarm intelligence distributed across the world wide web, self-reconfiguring robots, and others.

Quantitative, predictive understanding of complex systems requires comprehensive information. High-throughput methods and laboratory automation technology have the potential to deliver the necessary data. To harvest this potential, experimental design has to become self-evolvable and autonomous.

Autonomous experimentation systems are computational systems capable of autonomously investigating large experimental parameter space (Matsumaru et al. 2004; Lovel and Zauner 2009).

Such systems should develop hypotheses, plan experiments and perform experiments in a closed loop manner without human interaction.

In this new approach, artificial intelligence techniques are employed to carry out the entire cycle of cognition including the elaboration of hypothesis to explain observations, the design of experiments to test these hypotheses and the physical implementation of the experiments using laboratory automats to falsify hypotheses.

Table 8.2 Leading projects

Levels	Leading projects
Mat	Quantum Computer, Living Technologies, Self-fabrication
Bio	Genome Coding, Bio-Molecular Computer, Artificial Life
Cogno	Neural Coding, Neural Computer, Artificial Brain
Intel	Intelligence Coding, Autonomous Experiments, Artificial Intelligence, Smart Systems, Industry 4.0, Sustainable industry, Ecology
Math	Mathematical Coding, Automated Theorem Proving, Artificial Mathematics

In the coming decades a confluence of wireless networks and lab-on-chip sensor technology with application in health monitoring is expected. In such lab-on chip network each sensor node is endowed with a limited supply of chemicals. The network will collectively or via the self-evolution level decide how the drug resources will be spent. Environmental monitoring and improving, new drugs and new material discoveries may be performed by similar autonomous experimentation architectures (Symes et al. 2012).

8.3 Smart Society

8.3.1 Society 4.0

The digital revolution is not limited to industry but is changing all aspects of the human activity (Helbing 2014; 2015; Scharmer 2009; Scharmer and Kaufer 2013; Schwab 2016).

The primitive or pre-industrial society is considered as "society 0.0". The invention of the coal powered steam engines turned agricultural society ("society 1.0") into industrial society ("society 2.0"), and wide-spread education turned it into service society ("society 3.0"). The Internet, the World Wide Web, and Social Media are transforming service societies into digital societies ("society 4.0").

With computers reaching the level of human brainpower, with intelligent service robots, and the Big Data advent, the majority of jobs in the industrial and service sectors will be modified within the next years. Most of our current institutions will fundamentally change: the way we educate (Massively Open On-line Courses and personalized education), the way we do research (Big Data analytics), the transportation way (self-driving Google cars and drones), the way of shopping (Amazon and eBay), the way of producing (3D and 4D printers), but also our health system (personalized medicine), and most likely politics (participation of citizens) and the entire economy as well (the emerging sharing economy, and co-producing consumers). Financial business, which used to be the domain of banks, is increasingly replaced by algorithmic trading. How will this change the society is a challenging problem for complexity science.

Industry 4.0 must be understood as a future concept for society as a whole, the so-called "society 4.0" in which people, more than ever, are at the forefront. The increasing diversity of products with short delivery cycles and simultaneously decreasing numbers of personnel available can present an additional challenge for many companies. It is also important that the urban production of the future is moving closer to people.

This will require different logistics concepts for production supply and disposal. People are not being disregarded, quite the opposite in fact. Their requirements must be taken into account to a much greater extent in corporate planning in the future.

The industry 4.0 eco-system consists of smart factories and intelligent products with a memory that control production. It is a question of allowing people to perform high quality and creative work and giving them the opportunity to achieve a work/life balance—with just as much flexibility as the production systems of the future that will be controlled by people (Schwab 2016).

Each column of the Table 8.3 indicates the critical factors in each developmental stage (Scharmer and Kaufer 2013). The stages are denoted: 0.0, 1.0, 2.0, 3.0 and 4.0.

The lowest level 0.0 correspond to the natural, traditional or communal stage. The highest level 4.0 represents the vision fourth industrial revolution and society 4.0.

If we refer to coordination the community coordination 0.0, turned into hierarchical control 1.0, this into market competition 2.0, this into networks negotiation 3.0, and this will evolve into awareness based followed by collective actions 4.0.

Table 8.3 Society 4.0

	Agriculture	Capital	Computing technology	Consummation
0.0	Natural	Natural	Human	Survival
1.0	Subsistence	Human	Main frames	Traditional needs driven
2.0	Industrial	Industrial	PC	Consumerism mass consume
3.0	Selective cultures	Financial	Internet, IoT	Selective conscious
4.0	Collaborative conscious	Cultural Creation serving	Self-learning robots	Collaborative conscious
	Coordination	Defense	Design	Economy
0.0	Community	Individual	Individual	Tradition
1.0	Hierarchy control	Weapons	Traditional artifacts	Agriculture
2.0	Market competition	Machines	Products services	Industrial
3.0	Networks negotiation	Systems automation	Organizational transformation	Service
4.0	Awareness collective action	Unmanned self-organizing robot drones	Social transformation	Digital
	Education Learning	Energy	Food	Health
0.0	Tradition	Natural	Natural	Tradition
1.0	Memorization	Utilities	Cultivation	Examination dialogue
2.0	Open access library internet	Personal driven	Mechanization manufacturing	Open access internet
3.0	Knowledge producing	Renewable	Processing genetics	Semantic personalized
4.0	Innovation producing	Smart grids self-organized	Collaborative integrative	Symbiotic conscious

(continued)

Table 8.3. (continued)

	Industry	Labor Work	Leadership	Marketing
0.0	Manual	Self sufficiency	Community	Individual
1.0	Mechanic	Slave	Authority	Product centric
2.0	Electric	Labor commodity	Incentives	Customer Centric
3.0	Electronic automation	Labor regulated commodity	Participative	Human centric
4.0	Internet digital informational self-organized	Social business entrepreneur	Co-creative participation	Self-Actualization
	Mobility	Nature	Organization	Ownership
0.0	Individual	Mother Nature	Community	Communities
1.0	Public	Resources	Entrepreneur	State
2.0	Private cars	Commodity land materials	Professional	Private
3.0	Driver-less vehicles	Regulated commodity	Shared values	Mixed public/private
4.0	Autonomous transport systems	Eco-system commons	Project product facing-strategy	Shared common
	Social Stage	Technology	Web Internet	Work Security
0.0	Communal	Indigenous	Human direct	Individual
1.0	State centric	Tools mechanical	Read	Protect worker
2.0	Free market Ego-centric	Machines Industry	Social: read/write	Social security
3.0	Social market Regulated	System centric automation	Semantic: read/write/execute	Social market security
4.0	Co-creative Eco-centric	Human centric information	Symbiotic: read write/execute/concurrency	Self-organized Supply market

If we refer to economy, the traditional primitive economy 0.0, turned into agricultural economy 1.0, this into industrial economy 2.0, this into service economy 3.0, and this will evolve into digital economy 4.0.

If we refer to education, the traditions learning 0.0, turned into memorization learing 1.0, this into open access to libraries and education 2.0, this into knowledge production learning 3.0, and this will evolve into innovation production learning 4.0.

If we refer to industry we may associate industry 0.0 to manual energy, industry 1.0 to mechanical energy, industry 2.0 to electricity, industry 3.0 to electronics and automation and industry 4.0 to digitalization.

If we refer to web internet we may associate web 0.0 to texts and human direct interactions, web 1.0 to read capability, social web 2.0 to read/write, semantic web 3.0 to read/write/execute and symbiotic web 4.0 to read/write/execute/concurrency capability (Choudhury 2014).

In the 0.0 stages nature is frequently emphasized, indicating that this is the critical factor for the production function. Then, at stage 1.0, dependent labor became the critical developmental factor. The production function changes from one factor (nature) to two factors (nature, labor). In stage 2.0, when economies move from state-centered societies to market economies, industrial capital becomes the critical developmental factor. Capital allows the new players in the market economy to be productive, and as a result the production function of the economic system now has three factors (nature, labor, capital).

In the stage 3.0, technology emerges as a critical factor, and with that the factors of production evolve to four (nature, labor, capital, technology).

Finally, in the currently emerging stage 4.0, all of the factors may turn out to be bottlenecks, or critical factors, in the economy.

The stage 4.0 highlights the role of self-organization, cooperation, collaboration, co-evolution and connectivity. This stage is identified as "Self" in polytopic project implementations.

8.3.2 Society 5.0

Besides manufacturing processes or factory automation, cyber-physical have a broad spectrum of applications in domains such as education, healthcare, agriculture and so on. So far, a remarkable initiative is Society 5.0 in Japan (Keidanren 2016). Society 5.0 uses the technology as basis towards a human machine symbiosis throughout the society. It targets to improve the life of every person.

Super smart society or 'Society 5.0', is a concept introduced by the Japanese government under which various 21st century challenges are addressed: the ageing population, climate change, food security, the limited availability of natural resources and sustainable development.

Instead of the technology-driven approaches pursued so far, Japan's present strategy incorporates a human-centered approach. The goal is an inclusive society in which almost everyone shares the same opportunities and that guarantees economic growth and well-being for all.

The fundamental technologies of the concept are artificial intelligence, robotics and the Internet of Things. The construction and use of databases are to be designed in order to provide people with important new services. Intelligent transport systems, community care and healthcare systems, food production and distribution, and disaster prevention are examples of new research and development investment priority targets.

Mobility is a key to ensuring that people—including older—can live actively. More security and sustainability are also important in the transport system, so developing driverless cars is just as crucial as developing electric ones.

Table 8.4 compares the stages or basic development stages for different points of view.

Scharmer and Kaufer (2013).refer to stages, Helbing (2014) refer to economy and Keidanren (2016) to society.

Observe that the points of views are similar. The Keidanren approach uses the labels 1.0 to 5.0 were other initiatives or authors use the labels 0.0 to 4.0. In fact, there are 5 development stages.

Observe that comparable 5 development stages 0.0, 1.0, 2.0, 3.0 and 4.0 appear in the study of society 4.0 but also in the VDMA industry 4.0 toolbox (Anderl et al. 2015).

This illustrates the underlying common ground between different theories of development (Fischer 2008; Halford and Andrews 2010).

Japan's presented strategy incorporates a human-centered approach. On this basis, we could characterize society 5.0, as 5D reference architecture.

Selecting for instance four domains, from Table 8.4 each having 5 stages denoted by 0.0 to 4.0 we arrive at the 5D architecture (Table 8.5).

Figure 8.22 illustrates Society 5.0 Case-a.

The selected basic levels are Nature, Learning, Labor, and Economy.

The Case-a, refers to the societal choice in which based on natural resources, by learning and labor the economical inclusion is targeted (Table 8.5).

Figure 8.23 illustrates Society 5.0 Case-b.

The selected basic levels are Healthcare, Transport, Cities, and Marketing.

The Case-b, refers to the societal choice in which based on health care, transportation and city facilities the approach to the marketing is targeted.

Table 8.4 Stages comparison

	Stage	Economy	Keidanren
0.0	Communal	Primitive	Hunting 1.0
1.0	State centric	Agriculture	Agrarian 2.0
2.0	Free market Ego-centric	Industry	Industrial 3.0
3.0	Social market regulated	Service	Information 4.0
4.0	Co-creative Eco-centric	Digital	Super Smart 5.0

Table 8.5 Four domains-Case-a

	Nature	Learning	Labor	Economy
0.0	Mother Nature	Tradition	Self sufficiency	Tradition
1.0	Resources	Memorization	Slave	Agriculture
2.0	Commodity Land materials	Open access library internet	Labor Commodity	Industrial
3.0	Regulated commodity	Knowledge Producing	Labor regulated Commodity	Service
4.0	Eco-system commons	Innovation producing	Social business Entrepreneur	Digital

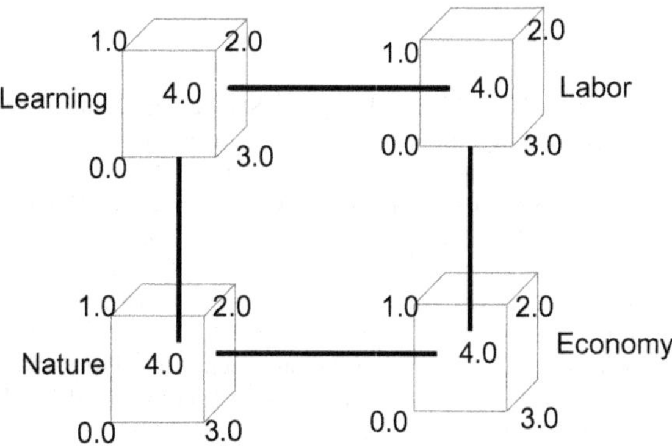

Fig. 8.22 Society 5.0 Case-a

Table 8.6 Four domains-Case-b

	Health	Transport	City	Marketing
0.0	Tradition	Individual	Tradition	Individual
1.0	Examination dialogue	Public transport	Private	Product centric
2.0	Open access internet	Private cars	City regulated	Customer Centric
3.0	Semantic personalized	Autonomous vehicles producing	Private/city regulated	Human centric
4.0	Symbiotic conscious	Uber transport systems	Smart city	Self-Actualization

Fig. 8.23 Society 5.0 Case-b

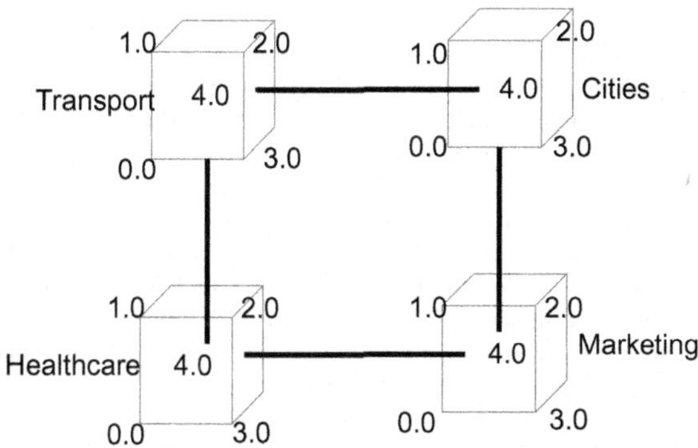

Observe that there are multiple aspects of Society 5.0, according to the interest of different members of the society. The individualized concepts of society 5.0 depend on age and on the general development of society.

Horizons beyond society 4.0 and society 5.0 towards high-dimensional society are waiting to be explored.

References

Anderl, R., Picard, A., Wang, Y., Fleischer, J., Dosch, S., Klee, B., Bauer, J.: Guideline Industry 4.0—Guiding principles for the implementation of Industry 4.0 in small and medium sized businesses. In: VDMA Forum Industry 4.0, Frankfurt (2015)

Baheti, R., Gill, H.: Cyber-physical systems. Impact Control Technol. **12**, 161–166 (2011)

Bainbridge, W.S., Roco, M.C.: Managing Nano-Bio-Info-Cogno Innovations: Converging Technologies in Society. Springer, Berlin (2006)

Bedau, M.A., McCaskill, J.S., Packard, N.H., Rasmussen, S.: Living technology: Exploiting life's principles in technology. Artif. Life **16**(1), 89–97 (2010)

Bryner, M.: Smart manufacturing: the next revolution. Chem. Eng. Prog. **108**(10), 4–12 (2012)

Choudhury, N.: World wide web and its journey from web 1.0 to web 4.0. Int. J. Comput. Sc. Inf. Tech. **5**(6), 8096–8100 (2014)

Davis, J., Edgar, T.F., Porter, J., Bernaden, J., Sarli, M.: Smart Manufacturing, Manufacturing Intelligence and Demand-Dynamic Performance. Comp. Chem. Engr. **47**, 134–144 (2012)

Fang, X., Misra, S., Xue, G., Yang, D.: Smart grid, the new and improved power grid: a survey. IEEE Commun. Surv. Tutorials **14**(4), 944–980 (2012)

Fischer, K.W.: Dynamic cycles of cognitive and brain development: measuring growth in mind, brain, and education. In: The Educated Brain: Essays in Neuroeducation, pp. 127–50 (2008)

Halford, G.S., Andrews, G.: Information—Processing Models of Cognitive Development. In: The Wiley-Blackwell Handbook of Childhood Cognitive Development, vol. 16, pp. 697–722 (2010)

Hartmann, N.: The New Ways of Ontology. Greenwood Press, Westport (1952)

Helbing, D.: 'What the digital revolution means to us'. Science Business, June 12 (2014)

Helbing, D.: Societal, economic, ethical and legal challenges of the digital revolution: from big data to deep learning, artificial intelligence, and manipulative technologies. Artificial Intelligence, and Manipulative Technologies, April 14, (2015)

Hwang, J.S.: The Fourth Industrial Revolution (Industry 4.0): Intelligent Manufacturing. SMT (Surface Mount Technology) Magazine, July, 10–15 (2016)

Iordache, O.: Self-Evolvable Systems. Machine Learning in Social Media. Springer, Berlin, Heidelberg (2012)

Iordache, O.: Polytope Projects Taylor & Francis. CRC Press, Boca Raton, USA (2013)

Iordache, O.: Implementing Polytope Projects for Smart Systems. Springer, Cham, Switzerland (2017)

Jorgenson, L.A., Newsome, W.T., Anderson, D.J., Bargmann, C.I., Brown, E.N., Deisseroth, K., Donoghue, J.P., Hudson, K.L., Ling, G.S., MacLeish, P.R., Marder, E.: The BRAIN Initiative: developing technology to catalyse neuroscience discovery. Phil. Trans. R. Soc. B **370**(1668), 20140164 (2015)

Kagermann, H., Lukas, W-D., Wahlster, W.: Industrie 4.0: Mit dem Internet der Dinge auf dem Weg zur 4. industriellen Revolution, VDI Nachrichten, **13**, (2011)

Kagermann, H., Wahlster, W., Helbig, J. (eds.): Recommendations for implementing the strategic initiative Industrie 4.0: Final report of the Industrie 4.0 Working Group (2013)

Keidanren: Reform of the economy and society by the deepening of "Society 5.0" Keidanren Japan Business Federation (2016)

Lee, J., Bagheri, B., Kao, H.A.: A cyber-physical systems architecture for industry 4.0-based manufacturing systems. Manufact. Lett. **3**, 18–23 (2015)

Lovel, C.J., Zauner, K-P.: Towards algorithms for autonomous experimentation. In: Naef, F., Galliot, B., Nehaniv, C., (eds.) Proceedings of the 8th International Conference on Information Processing in Cells and Tissues (IPCAT 2009), pp. 150–152 (2009)

Matsumaru, N., Centler, F., Zauner, K. P., Dittrich, P.: Self-adaptive Scouting-Autonomous Experimentation for Systems Biology. In: Lecture Notes in Artificial Intelligence, **3005**, Lecture Notes in Computer Science, pp. 52–62, Springer, Berlin (2004)

OMG.: Object Management Group, Software & Systems Process Engineering Meta-Model Specification 2.0 (2008)

Rasmussen, S.: The "4th Industrial Revolution" in fact a More Radical "BINC Revolution"?. In: Conference American Chamber of Commerce, China, Beijing, December 14, (2016)

Scharmer, O.: Theory U: Leading from the Future as It Emerges. Berrett-Koehler, San Francisco (2009)

Scharmer, O., Kaufer, K.: Leading from the Emerging Future: From Ego-System to Eco-System Economies. Berrett-Koehler, San Francisco (2013)

Schwab, K.: The Fourth Industrial Revolution. World Economic Forum, Geneva (2016)

Sharp, P.A., Langer, R.: Promoting Convergence in Biomedical Science. Science **333**, 527 (2011)

Symes, M.D., Kitson, P.J., Yan, J., Richmond, C.J., Cooper, G.J., Bowman, R.W., Vilbrandt, T., Cronin, L.: Integrated 3D-printed reactionware for chemical synthesis and analysis. Nat. Chem. **4**(5), 349–354 (2012)

Thoben, K. D., Wiesner, S., Wuest, T.: Industrie 4.0 and Smart Manufacturing—A Review of Research Issues and Application Examples. Int. J. Autom. Technol. **11**(1), 4–16 (2017)

Uslar, M., Engel, D.: Towards generic domain reference designation: how to learn from smart grid interoperability. Poster Proc. DACH Energy Inform. **2015**, 1–12 (2015)

Uslar, M., Gottschalk, M.: Extending the SGAM for Electric Vehicles. In: International ETG Congress 2015; Die Energiewende-Blueprints for the new energy age; Proceedings of 2015 Nov 17, pp. 1–8 (2015)

Uslar, M., Trefke, J.: Applying the smart grid architecture model SGAM to the EV Domain. In: EnviroInfo 2014, pp. 821–826. (2014)

von Kiedrowski, G., Otto, S., Herdewijn, P.: Welcome home, systems chemists! J. Syst. Chem. **1** (1), 1 (2010)

Wang, Y., Towara, T., Anderl, R.: Topological Approach for mapping technologies in reference architectural model Industrie 4.0 (RAMI 4.0). In: Proceedings of the World Congress on Engineering and Computer Science 2017, vol. 2. (2017)

Zuehlke, D.: Smartfactory–From vision to reality in factory technologies. IFAC Proc. **41**(2), 14101–14108 (2008)

Advanced Polytopic Projects

Octavian Iordache

Polytopic projects are based on a biologically inspired general framework shared by the functional organization of organisms as informational and cognitive systems, the scientific and engineering methods and the operational structure of smart devices, technologies, organisms or societies. The proposed reference architecture supports the investigation, fabrication and implementation of industrial systems required by the 4th industrial revolution advent.

The 4D approach studied in previous monographs is extended here to 5D and 6D approaches. This natural advancement is of great industrial and scientific interest.

The book is divided in 8 chapters. Chapter 1 outlines the limits of split in analytic and synthetic approaches and justifies the need of specific coupling of both epistemological ways in technology and science. The 4D, 5D and 6D polytopic architectures are proposed as a basic guide and reference architecture, for understanding and solving problems, for designing building and controlling of self evolvable and self-sustainable smart systems.

Chapter 2 is dedicated to integration and separation methods for technological schemes. Rooted trees and polytopic separations schemes allowing multi-scale processing are presented. Structuring and restructuring in supramolecular chemistry and quasi-species theory are presented in Chap. 3. This includes 6D polytopes. Chapter 4 outlines conditioning and randomness significance for processes. Forward and backward evolution is described and exemplified for mixing and restricted transfer processes. The 5D and 6D transfer devices are proposed. Chapter 5 examines assimilation and accommodation aspects for cognitive architectures. The role of development stages for cognition is outlined. Relation with dynamic skill theories is emphasized by a development polytope. Chapter 6 focuses on testing and designing relation to data. The pharmaceutical polytope, learning polytope and elements of design hermeneutics are introduced. Chapter 7 presents industrial systems based on additive and subtractive technologies. The role of

O. Iordache, *Advanced Polytopic Projects*, Lecture Notes
in Intelligent Transportation and Infrastructure,
https://doi.org/10.1007/978-3-030-01243-4

signed graphs is emphasized. High-dimensional printing and modular robots self-reconfigurations are presented.

Chapter 8 evaluates the perspectives and contains a retrospective of the domain for polytopic projects as physical, informational and conceptual systems. Smart enterprises, smart factory, generic reference architectures, smart technologies and smart society 4.0 and 5.0 projects are discussed.

Chemical engineering, chemistry, pharmaceutics, material science, and systems chemistry are the preferred domains for examples highlighting the power of the polytopic projects to achieve smartness in high complexity conditions. The book will be useful to engineers and scientists working in chemical engineering, chemistry, biochemistry, pharmaceutics, material science, systems chemistry, environment protection and ecology, to entrepreneurs and students in different domains of complex systems production and engineering, and to applied mathematicians.

Appendix A
Dual Graded Graphs

A significant generalization of the concept of differential posets [Stanley, R.: Differential posets, J. Amer. Math. Soc. **1**, 919–961 (1988)] is that of dual graded graphs, DGG [Fomin, S.: Duality of graded graphs, J. Algebraic Combin. **3**, 357–404 (1994)]

A graph is said to be graded if its vertices are divided into levels numbered by integers, so that the endpoints of any edge lie on consecutive levels.

A graded graph is a triple G = (P, ρ, E) where:

- P is a discrete set of vertices
- ρ: P → Z is a rank function
- E is a multiset of arcs (x, y) where ρ(y) = ρ(x) + 1

The set P_n = {x: ρ(x) = n ∈Z} are called levels of G.

G can be regarded either as oriented or as non-oriented graph. The accessibility relation in an oriented graph defines a partial order on P. If there are no multiple arcs, G is the Hasse diagram of this poset. The paths in non-oriented graphs are the Hasse walks.

As for the differential posets the "up" U and "down" D operators can be defined.

Let G = (P, ρ, E) be a graded graph. Linear operators U and D are defined by:

$$Ux = \sum_{(x,y)\in E} m(x,y)y, \quad Dy = \sum_{(x,y)\in E} m(x,y)x \tag{A.1}$$

Here m (x, y) is the multiplicity of the edge (x, y) in E.

A significant idea was to consider the pairs of dual graded graphs G_1 = (P, ρ, E_1) and G_2 = (P, ρ, E_2) with a common set of vertices and a common rank function.

The oriented graded graph G = (G_1, G_2) = (P, ρ, E_1, E_2) is then the directed graded graph on P with edge in E_1 directed upwards, and edges in E_2 directed downwards.

The "up" U and "down" D operators associated with the graph G = (G_1, G_2) are defined by:

O. Iordache, *Advanced Polytopic Projects*, Lecture Notes in Intelligent Transportation and Infrastructure,
https://doi.org/10.1007/978-3-030-01243-4

$$Ux = \sum_{(x,y)\in E_1} m_1(x,y)y, \quad Dy = \sum_{(x,y)\in E_2} m_2(x,y)x \tag{A.2}$$

Here $m_i(x, y)$ denotes the multiplicity of (x, y) in E_i.

Let (G_1, G_2) be an oriented graded graph such that:

- It has a zero,$\hat{0}$
- Each rank has a finite number of elements

The graphs G_1 and G_2 are said to be dual as operators in $G = (G_1, G_2)$ if:

$$D_{i+1}U_i - U_{i-1}D_i = r_i I_i \tag{A.3}$$

Here U_i (respectively D_i) denote the restriction of the operator U (respectively D) to the ith level of the graph, I_i denote the identical operator at the same level and r_i are positive integers. If $r_i = r$, G_1, and G_2 are r-dual graphs and we call the pair (G_1, G_2) an r-dual graded graph.

The 1-dual graphs satisfy the relation:

$$DU - UD = I \tag{A.4}$$

The quantum dual graphs satisfy the relation:

$$DU - qUD = I \tag{A.5}$$

They have been introduced by Lam, T.: Quantized dual graded graphs. Electron. J. Combin. **17**(1), Research Paper 88, (2010)

The dual filtered graphs satisfy the relation:

$$DU - UD = D + I \tag{A.6}$$

They have been introduced by Patrias, R., Pylyavskyy, P.: Dual Filtered Graphs. arXiv preprint arXiv:1410.7683 (2014)

Appendix B
Hopf Algebras

The aim of this appendix is to introduce basic concepts of the bialgebra and Hopf algebra (Dăscălescu, S., Năstăsescu, C., Raianu, Ș.: Hopf Algebras. An introduction, Pure and Applied Mathematics, 235, Marcel Dekker 2001).

Let k be a commutative ring, one calls algebra over k, or also k-algebra, a module A together with a linear mapping: $A \otimes A \rightarrow A$.

Let A be algebra over k.

The algebra A is of finite dimension over k if it is a finite dimensional as a vector space.

The algebra A is associative if the inner product is so.

A is an algebra with unit if the inner product has a neutral element.

Let A be an associative algebra A with unit. An element of A is invertible if it is invertible under the ring structure.

The algebra A is commutative if the inner product is so.

One defines an associative algebra with unit as follows: an associative algebra over k is a triple (A, ∇, η), where A is a k-module, $\nabla: A \otimes A \rightarrow A$ and $\eta: k \rightarrow A$ are morphisms of k-vector spaces such that: $\nabla \circ (\eta \otimes id_A)$ and $\nabla \circ (id_A \otimes \eta)$ are scalar multiplication.

Coalgebras are structures that are dual to unital associative algebras.

The axioms of unital associative algebras can be formulated in terms of commutative diagrams. Turning all arrows round, one obtains the axioms of coalgebras.

Every coalgebra, by duality, gives rise to an algebra, but not in general the other way.

In finite dimension, this duality goes in both directions.

Formally, a coalgebra over a field k is a vector space C over k together with k-linear maps $\Delta: C \rightarrow C \otimes_k C$ and $\varepsilon: C \rightarrow K$ such that:

$$(\mathrm{id}_C \otimes \Delta) \circ \Delta = (\Delta \otimes \mathrm{id}_C) \circ \Delta \tag{B.1}$$

$$(id_C \otimes \varepsilon) \circ \Delta = id_C = (\varepsilon \otimes id_C) \circ \Delta \tag{B.2}$$

O. Iordache, *Advanced Polytopic Projects*, Lecture Notes in Intelligent Transportation and Infrastructure,
https://doi.org/10.1007/978-3-030-01243-4

A bialgebra over a field k is a vector space over k which is both a unital associative algebra and coalgebras, such that these structures are compatible.

$(B, \nabla, \eta, \Delta, \varepsilon)$ is a bialgebra over k if it has the following properties:

1. B is a vector space over k
2. There are k-linear maps (multiplication or product) $\nabla: B \otimes B \rightarrow B$ and unit $\eta: k \rightarrow B$, such that (B, ∇, η) is a unital associative algebra
3. There are k-linear maps (comultiplication or coproduct) $\Delta: B \rightarrow B \otimes B$ and counit $\varepsilon: B \rightarrow k$, such that (B, Δ, ε) is (counital coassociative) coalgebras
4. Compatibility conditions are expressed by the following commutative diagrams:

Figure B.1 refers to product and coproduct.

Figure B.2 refers to product and counit, coproduct and unit, unit and counit.

Here $\tau: B \otimes B \rightarrow B \otimes B$ is the linear mapping defined by $\tau(x \otimes y) = y \otimes x$ for all $x, y \in B$. The diagrams (∇, Δ) and (∇, ε) express that the multiplication ∇ is

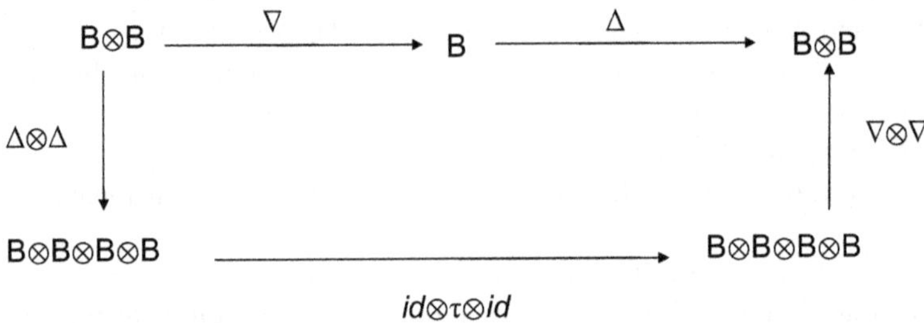

Fig. B.1 Product and coproduct

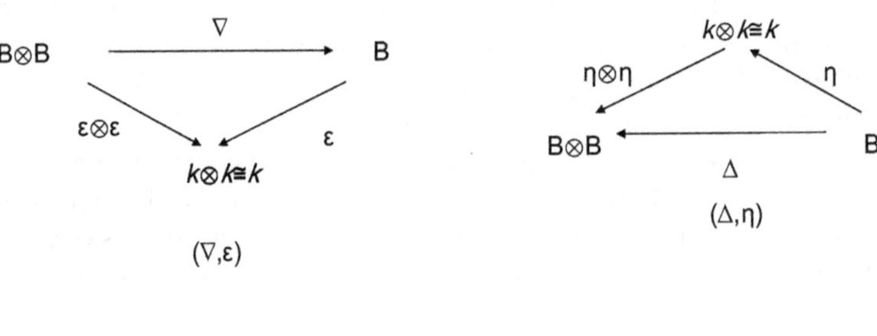

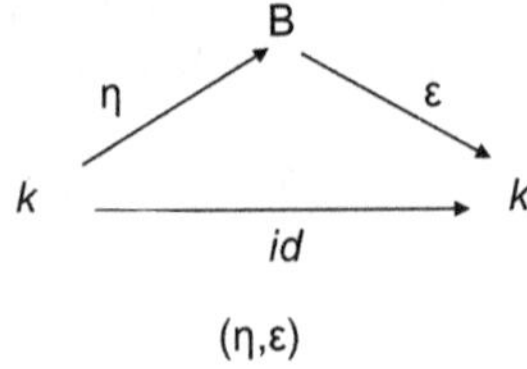

Fig. B.2 Product and counit, coproduct and unit, unit and counit

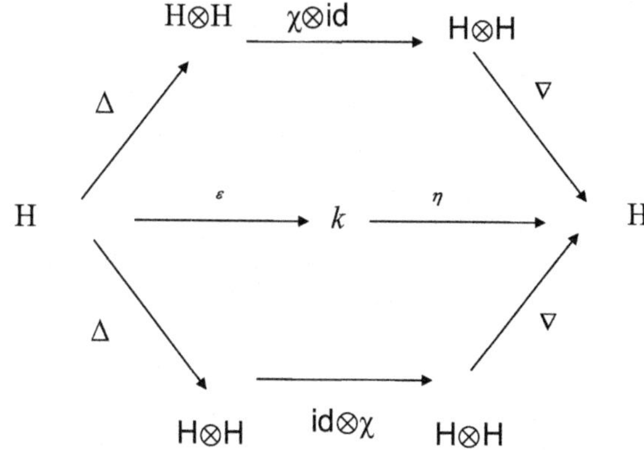

Fig. B.3 Hopf algebra diagram

a morphism of coalgebras. The diagrams (Δ, η) and (η, ε) express that the unit η is a morphism of coalgebras. The diagrams (∇, Δ) and (Δ, η) express that the comultiplication Δ is a morphism of algebras. The diagrams (Δ, ε) and (η, ε) express that the counit ε is a morphism of algebras.

A bialgebra is a *k*-vector space B, endowed with an algebra structure

(B, ∇, η), and with a coalgebras structure (B, Δ, ε) such that ∇ and η are morphism of coalgebras; equivalently, it follows that Δ and ε are morphism of algebras.

A Hopf algebra is a (associative and coassociative) bialgebra H over a field *k* together with a *k*-linear mapping χ: H → H, called the antipode, such that the diagram from Fig. B.3 commutes. Figure B.3 shows the Hopf algebra diagram.

The antipode χ sometimes required to have a *k*-linear inverse, which is automatic in the finite-dimensional case, or if H is commutative or cocommutative (or more generally quasi-triangular). If $\chi 2 = id_H$ then the Hopf algebra is said to be involutive.

If H is finite-dimensional semisimple over a field of characteristic zero, commutative, or cocommutative, then it is involutive. If a bialgebra B admits an antipode χ it is unique. The antipode can be equivalently defined as the inverse of id_H for the associative convolution product.

Define a space of homomorphisms from co-algebra (B, ∇, η) to algebra (B, Δ, ε), denoted by Hom (B^c, B^a) with respect to a convolution product denoted by $*$ and defined as:

$$f * g = \nabla \circ (f \otimes g) \circ \Delta, \quad \forall f, g \in \mathrm{Hom}(B^c, B^a) \tag{B.3}$$

The diagram from Fig. B.3 can be written:

$$\chi * id_H = id_H * \chi = \eta \circ \varepsilon \tag{B.4}$$

Index

O. Iordache, *Advanced Polytopic Projects*, Lecture Notes in Intelligent Transportation and Infrastructure,
https://doi.org/10.1007/978-3-030-01243-4

Printed in the United States
By Bookmasters